*This book is dedicated
to my parents,
Alexandra Wujec,
who showed me the fun of discipline,
and Stanley Wujec,
who showed me the value of letting go.*

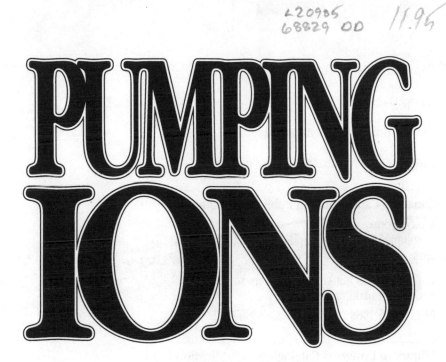

PUMPING IONS

Doubleday Canada Limited, Toronto
Doubleday and Company, Inc., Garden City, New York

DESIGN: Brant Cowie / Artplus Ltd.

PAGE MAKE-UP AND ILLUSTRATION: Tom Wujec

Canadian Cataloguing in Publication Data
Wujec, Tom.
  Pumping ions: games and exercises to flex your mind
ISBN 0-385-25113-0
1. Thought and thinking. 2. Problem solving.
3. Decision-making. I. Title.
BF455.W85 1988   153.4'2   C87-094096-1

Library of Congress Cataloging-in-Publication Data
Wujec, Tom.
  Pumping ions.
  1. Thought and thinking — Problems, exercises, etc.
I. Title.
BF441.W84 1988   153.4'2   87-36507
ISBN 0-385-24749-4

Printed and bound in Canada by Gagne Ltd.

Published in Canada by
  Doubleday Canada Limited
  105 Bond Street
  Toronto, Ontario
  M5B 1Y3

Published in the United States by
  Doubleday and Company Inc.
  666 5th Avenue
  New York, N.Y.
  10103

Table of Contents

Preface

I would like to thank Doubleday for their support, encouragement, and enthusiasm for this project. *Pumping Ions* was written, illustrated, and typeset entirely on computers. The manuscript was written on WordPerfect 4.2 on a Toshiba T3100 and various other computers. Page makeup was performed using Ventura Publisher 1.1. The illustrations were produced on a Macintosh SE, using Adobe Illustrator 1.0, and on an IBM AT compatible using the Artwork Environment and Publisher's Paintbrush. *Pumping Ions* was set in Garamond Light. The final type was modified using custom Postscript procedures. Proofs were printed on an Apple LaserWriter Plus and final pages were printed on a Linotronic 100.

I would also like to thank several people who helped out on *Pumping Ions*: Peter Taylor, my agent, who saw its potential; Charlie Menendez for all those inspiring conversations that went into the wee hours of the morning; Brant Cowie for his design ideas; Maggie Reeves for keeping the production on track and on time; Pippa Campsie, who offered suggestions, ideas, and gave great lunches; Peter Turney, who never failed to stimulate my brain; Barney Gilmore for generally being an all-round inspiration; Chris Sasaki for his insights; Ken Deaton for his real-time brainstorming; Wayne and Pam Weaver of MainFrame Computer Graphics, and Patrick Lee of Interaxis, for their technical wizardry; Larry and Ellen Oberlander for showing me that there was a gymnasium; and E.J. for showing me what it means to work out.

Most of all, I would like to thank my wife, Susan, for her thoughts and suggestions, and for bearing through the long months of writing. I can never thank her enough for her patience, love, and support.

Tom Wujec
Valentine's Day, 1988
Toronto, Canada

MACK'S MENTAL GYMNASIUM

Mack's Mental Gymnasium

Tom's mind felt like a bowl of vanilla pudding. He'd been in meetings all day and now felt distracted and preoccupied, a feeling that was becoming all too familiar. Recently his attention span seemed to be getting shorter, and his ideas were increasingly mediocre. Tom realized that he was thinking about the same old things in the same old ways.

As he paced the familiar path back from work, he noticed a new shop sign:

Interesting. On an impulse, Tom opened the door and went in.

Upstairs, Tom met Mack, an animated fellow with a sparkle in his eye, who was the owner as well as the fitness coordinator. Tom explained, "I just don't seem to have the mental energy and flexibility I used to have."

Mack thought for a moment and then asked Tom several questions. *"Have you really challenged your mind lately?"*

"Well, not that I can remember," Tom answered.

"How often do you do nothing and relax?"

"I watch some TV. It helps me unwind."

"Have you been learning anything new?"

"You see, I just don't seem to have the time."

"What kind of food do you feed your mind?"

"Well, I read a newspaper every now and then."

The conversation went on like this for a while. Mack asked questions about paying attention to everyday things, about searching for problems and opportunities, about pursuing long-term goals, and about learning for learning's sake. Finally Tom asked, "Well, where do I stand?"

"There's no doubt about it," Mack announced. "You're mentally out of shape.

"You see, in some ways, your mind is like your body. If it doesn't get the exercise it needs, it becomes stiff and weak. If your brain sits about idly, without ever working up a sweat, your mental muscles become sluggish. This lack of conditioning can lead to hardening of the attitudes, poor circulation of ideas, gain in mental flab, excessive tension, boredom, and, worst of all, mental constipation."

"How did I get this way?" Tom wanted to know.

Mack explained, "People get mentally out of shape when they stop challenging their minds. This happens when you become mentally complacent and opt for quick, habitual solutions, rather than purposeful thought. It also happens when you confine thinking to a small range of interests. For example, some people are strong at designing bridges and making good business decisions, but weak at searching out creative ideas, or organizing their time, or holding a good conversation, or telling a joke, or relaxing their minds. They use their brain well in some ways, but not in others.

"When people stop having fun using their minds to search out new ideas, by experimenting, by playing around with new possibilities, they become mentally rigid. They forget that, to a large degree, their world is created by their minds. They're so busy focusing outside of themselves that their insides suffer. Problems, worries, and responsibilities take the forefront and they forget to pause every once in a while and think about *how* they think. A healthy mind is able to move in different ways. To be at your best, you need to exercise all of your mental muscles, and only a well-rounded workout will give you this."

Tom was astonished. What Mack was saying made so much sense, and it was so simple. He said, "I can see that I've been letting some of my mental muscles sit idly. And when I do use them, I use them in the same old ways. Is it too late? Can I still get fit?"

"It's never too late to exercise," Mack replied. "Ideally everyday circumstances should give us the challenges to stretch and flex all our mental muscles. But if we get caught up in routine, or if we get lazy, or if we don't have a good set of techniques for reaching our potential, we need to pay a visit to the mental gymnasium."

Mack turned around and led Tom through a small doorway to a spacious, brightly lit room: Mack's mental gymnasium. On the walls were anatomical illustrations of various mental muscles — reasoning muscles, concentration muscles, visualization muscles — along with techniques to exercise them. Scattered throughout the room were several exercise stations. At each station, people were engaged in different activities: talking, drawing, writing, laughing, concentrating. Tom figured out that people moved from one station to the next when they wanted to exercise a different part of their minds.

Mack continued, "It takes time, patience, and determination to eliminate old habits and replace them with new habits. But the energy you put into improving your thinking always pays you back with improved creativity and productivity. Besides, as you learn how your mind works (and how it doesn't), you'll have a lot of fun."

Tom felt as if he were about to embark on a long, invigorating adventure. As he watched the people mentally jogging, power lifting, and stretching, all of them obviously enjoying themselves, he realized that he had found something he didn't consciously know he'd been searching for: a place to really challenge and improve his mind. He turned to Mack with a smile and asked, "When can we start?"

Mack gave the answer he always gave. "Right now, of course."

Better Thinking

and

Thinking Better

The Anatomy of Mental Fitness

"Strength of mind is exercise, not rest."
ALEXANDER POPE, *18th-century poet*

EXERCISE How would you rate your level of mental fitness?

❑   **HIGH** My mental performance is exceptional.
Just call me Leonardo.

❑   **MEDIUM** I occasionally think of new things
and sometimes challenge my thinking muscles.

❑   **LOW** My mind is like a bowl of tapioca pudding.
I'm in real need of a mental workout.

No matter which category you selected, remember that mental fitness — your ability to concentrate, to reason, to visualize, to imagine, to make decisions, to solve problems, and to think clearly and creatively — depends greatly on how well and how often you exercise your mind. If you feel that you are mentally out of shape, then cheer up; you can improve simply by taking the time to work out your mental muscles. If you feel fit, then recognize that you still need exercise to keep your mind in top shape. Even Olympic athletes need practice.

Making your thinking muscles stronger involves asking yourself:

How can I improve my mental performance?

First become familiar with the different mental muscles within your mind. To do this, try the following exercise.

Mental Tour

Imagine that you're holding an orange.
Picture what it would feel like,
what it would look like,
and what it would smell like.

For a few moments,
try to form as clear an image as you can.

Now imagine peeling off the skin,
pulling apart the segments,
and biting into a piece.

After a moment, examine a segment very closely.
Ask yourself what the orange would look like
if you could enlarge it a thousand times, or a million times.
What might the cells look like?
What might the molecules look like?

For the next couple of minutes,
try to become aware of everything
that you know and don't know about oranges.

Think about what makes an orange an orange,
why it tastes the way it does,
how many types of oranges there may be,
how oranges may have evolved through time,
what purposes oranges may serve,
and how to make a good orange marmalade.

As you think about the orange,
pay close attention to the quality of your thoughts.
Put this book down for a moment and start thinking.

If you're like most people, you probably noticed that the longer you thought about the orange, the more ideas, associations, and connections popped into your imagination. You may have thought about the physical properties of oranges. You may have pondered the history and economics of oranges. You may even have considered the fact that there is no English word that rhymes with "orange". As you formed mental pictures, reached back into your memory, and wondered, you shifted between one mode of thought to another: you were moving different mental muscles.

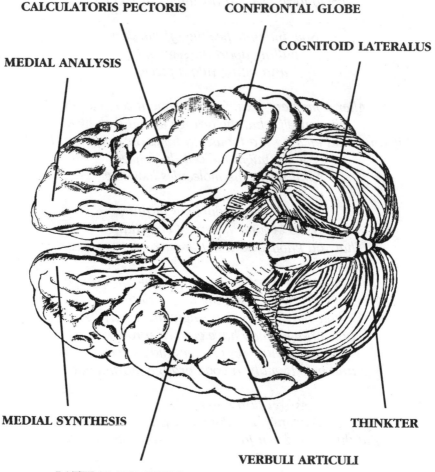

CALCULATORIS PECTORIS   CONFRONTAL GLOBE

COGNITOID LATERALUS

MEDIAL ANALYSIS

MEDIAL SYNTHESIS   THINKTER

VERBULI ARTICULI

LATERAL IMAGINUS

There are mental muscles for each kind of thinking in which we engage. Logical thinking, metaphorical thinking, analytical thinking, critical thinking, verbal thinking, and visual thinking — each represents just one of the mental muscles that enable us to move through our inner world.

In the gymnasium of life, you need to flex your mental muscles in different ways depending on the needs at hand. Sometimes you push hard by thinking critically and dealing with cold, hard logic, and at other times, you relax, let go, and playfully stretch out in new directions. You may work diligently at a steady pace and get lots accomplished, or you may balance several factors and arrive at elegant solutions to problems. In the same way that several body muscles work together to create physical movement, several mental muscles work together to create clear, purposeful thinking. In a nutshell, you could say that there are four basic qualities that characterize a fit mind. These qualities are:

Mental Strength

Mental Flexibility

Mental Endurance

Mental Coordination

Each time you apply your mind to a task that exacts concentration, you apply your *mental strength*. You use it when you sort through the options in a difficult decision, work through a complex math problem, balance a checkbook, or narrow your attention and focus it on one thing or idea and keep it there. Mental strength is the ability to concentrate when you want to, as hard as you want to, for as long as you want to.

During the times when you need to be innovative and creative, your thinking muscles need to be pliant and supple. *Mental flexibility* is your capacity to switch from one mode of thought to another. It's a playful process as you toy about, find new combinations, and look in new directions. You bend concepts, twist ideas, and put your mind in unusual postures to explore new possibilities. Mental flexibility is artistry, holistic thought, creativity, brainstorming, and a bit of Zen, all rolled into one.

When you want to put your ideas into action, you need *mental endurance*. Mental staying power is the ability to sustain an increased level of activity without getting distracted or discouraged. It's the capacity to persist, to go the full mile.

And when you want to add precision and flair, you need *mental coordination*. Mental coordination is timing, balance, and agility. It's the knack of gracefully orchestrating your thinking so that you can deal with several things at once, keep balanced within the face of uncertainty, learn for knowledge's sake, and strive for higher ideals.

It's these four basic qualities — strength, flexibility, endurance, and coordination — that make up a fit mind. Only by stretching, flexing, exerting, relaxing, moving your mind in different ways, performing different mental movements, does your mind get in shape.

"As diamond cuts diamond,
and one hone smooths a second,
all the parts of intellect are whetstones to each other;
and genius is the result of their mutual sharpening."
CYRUS BARTOL, *19th-century clergyman*

How Do People Get Mentally Out of Shape?

Why are some people's minds sharp, energetic, and overflowing with creative ideas and other people's intellects less than dazzling? There are two main reasons: the demands made by circumstances and the pull of mental habits.

A stonemason doesn't need to go to the gym to build up his arms. In the course of his work of mixing mortar and hauling bricks, his arms are naturally strengthened. Similarly, an accountant doesn't need to go to school to tighten up his arithmetic skills. Because he is constantly dealing with numbers, his mathematics muscles are kept trim.

If your circumstances don't challenge your thinking muscles, you don't get the stimulation you need to keep fit. However, if you are often put in situations where you need to think clearly and energetically, your muscles become stronger and more responsive. Perhaps this idea is summarized best in the old saying:

TIP Ask yourself: Which of my mental muscles don't get a daily workout?

A habit is simply the tendency to do something without having to think about it. Like it or not, your whole life, from the way you brush your teeth to the way you strive towards long-term goals, depends largely on your personal repertoire of automatic tendencies. It has to. Could you imagine having to relearn how to brush your teeth every day?

Your mental life — what you notice, how much you pay attention, how well you learn, how you approach and deal with problems, what you remember, what you worry about, what you enjoy, what you think about all day long — is also largely guided by habits.

The key to mental fitness is to develop a good set of habits that help you get to where you want to go. Mentally fit people can exert themselves whenever they want to. They are interested in how the world works and why things happen the way they do. They enjoy a wide range of interests. They also know how to change habits through conscious, deliberate action and can form effective tendencies, such as the habit to daydream less, or to weigh decisions more carefully, or to take more risks. Together, these good habits encourage the mind to become more responsive and ultimately to develop the most important habit of all: the habit to develop good habits.

> *"Sow a thought, and you reap an act.*
> *Sow an act, and you reap a habit.*
> *Sow a habit, and you reap a character.*
> *Sow a character, and you reap a destiny."*
>
> CHARLES READE, *19th-century novelist*

TIP Ask yourself: Which of my mental habits weigh me down?

What Makes a Good Mental Exercise?

Solving a crossword puzzle — Preparing for an exam — Conducting a television interview — Improvising on stage — Studying a difficult text — Cooking a gourmet dinner — Making up a chewing gum jingle — Visualizing a friend's face — Recalling the last time you had ice cream — Reciting the national anthem backwards — Naming your elementary school teachers — Continuing the sequence: 2, 4, 8, 16, as long as you can — Learning a new language — Telling a convincing lie — Programming a computer — Figuring out how a toaster works — Drawing a realistic landscape — Changing a bad mood into a good mood — Contemplating infinity — Writing a docudrama — Recalling an important conversation you had last month — Asking your boss for a raise — Finger painting — Controlling your temper — Reorganizing your bedroom — Questioning authority — Designing a better mousetrap — Explaining why the sky is blue to a four-year-old — Negotiating a contract — Bluffing in a poker game — Playing a video game — Searching for truth

How do you get mentally in shape? Mental exercise, like physical exercise, involves movement. For your body, movement is running, swimming, playing basketball, lifting weights, or stretching — any activity that makes your muscles expand and contract. For your mind, movement is progression of thought, a kind of inner travel from premise to conclusion, from problem to solution, from question to answer, from answer to question, from one state of mind to another state of mind. Whenever you actively apply your mind to some task and intentionally manipulate your mental resources, you move your mental muscles.

In the broad definition of the term, an exercise is an activity performed for the purpose of improving a skill, bettering yourself, or training your faculties. Virtually any task that requires *active attention* — such as figuring out a puzzle, working through a business problem, or sitting quietly and collecting your thoughts — can be a mental exercise.

Perhaps more important than *what* you do is *how* you do it — good technique is essential. A gymnast improves by repeating a move over and over, making slight adjustments, until it's just right. Through this repetition and feedback, the gymnast's muscles become more responsive. Similarly, to improve your mental muscles, you practice thinking. In your mind, you repeat a task until you get your thinking muscles to do what you want them to do.

A good mental exercise is an appointment with yourself, a time to turn your focus inward and challenge your mind. It's a way to channel your mental energy into constructive thought. Whether you want to work hard to reach peak mental performance, or to stretch out a bit to become more flexible, good old-fashioned mental exertion can enhance your creativity, give you a sense of accomplishment, and improve your state of mind.

Creating Your Own
Personal Gymnasium

In the following twelve chapters, you'll explore a multitude of ways to work out your brain. Each chapter is an exercise station that presents a variety of mental exercises. At some stations you'll stretch, relax, and sink into peaceful tranquillity. At other stations you'll work your inner muscles with drills and mental calisthenics until your mind starts to sweat. While some exercises work your left brain — the analytic, logical part of your mind — others work your right brain — the intuitive, spatial part of your mind. Together, they give you a well-rounded mental workout.

Station One: Loosening Up *Collecting Your Attention*	**Station Seven:** Mental Play *Fooling Around in Your Head*
Station Two: Mental Movement *Concentration*	**Station Eight:** Reaching Back *Memory Muscles*
Station Three: Mental Calisthenics *Increasing Your Endurance*	**Station Nine:** Mental Flexibility *Analysis and Synthesis*
Station Four: Mental Gymnastics (I) *Thinking with Images*	**Station Ten:** Mental Balance *Decision Making*
Station Five: Mental Gymnastics (II) *Thinking with Words*	**Station Eleven:** Improvising *The Creative Act*
Station Six: Mental Strength *Problem Solving*	**Station Twelve:** Peak Performance *Learning About Learning*

Exercise Style Tips

Here are some rules of thumb to improve your exercise routine.

TIP ONE Change out of your mental street clothes. Leave your problems and concerns outside in the locker room. Adopt the workout attitude by intending to strengthen yourself. Delight in the difficulties that you choose for yourself.

TIP TWO Exercise — don't analyze. Since the whole point of mental exercise is consciously to manipulate your mental resources, *do* the exercises. The amount of benefit you receive is directly proportional to how much you are willing to stretch and exert your mental muscles. Remember, good form is the key to good exercise.

TIP THREE Take your time. Don't rush your workout. It takes a while to enter deeper layers of thinking, so be patient. Give yourself plenty of time to explore your inner world.

TIP FOUR Repeat the exercises. A lot of the exercises in this book can be performed many times on many separate occasions. Only with time and practice will your abilities improve. Make up a routine of your favorite mental calisthenics. If you reach a block, an exercise you can't do now, take a break and come back to it later.

TIP FIVE Exercise often. The more you exercise your mind, the easier it becomes. The easier it becomes, the more you enjoy it. The more you enjoy it, the more you exercise your mind. And the more you exercise your mind, the more fit your mind becomes.

Now that you're ready, let's get moving.

1

LOOSENING UP

Collecting Your Attention

Start Now

Don't move any part of your body
from the position that it's in right now.
Examine your posture,
the expression on your face,
the position of your fingers.

Search for any tightness in your jaw,
in your eyebrows,
in your stomach,
and in your legs.

Are your shoulders raised?
Are your toes curled?
Is your body leaning?
If you relaxed your muscles totally,
in what direction would you fall?

Get a clear sense of your body and mind.
Relax any tensions that you may discover,
take a couple of slow deep breaths,
then continue.

The Play of Attention

"A quiet mind cureth all."

ROBERT BURTON, *17th-century philosopher*

Imagine that you have at your disposal a total of one hundred volts of attention. Every time you put your mind to something, you engage a portion of this mental energy. However, whenever you become distracted or preoccupied, some of this energy becomes wasted. Say that you're working at your desk on a project. As you're working, imagine that your body settles into a slightly uncomfortable posture — some of your muscles become tense. When this happens, like it or not, part of your attention is drained away. Though you may not be aware of the physical tension, it has an effect on your mental energy. You lose, say, twenty-five volts of attention, a quarter of your full supply.

Now imagine that, as you continue to work at your desk, you feel a little bored. You begin to look at your project as a chore, and you feel an inkling of conflict. Part of you wants to work and part of you doesn't. Another twenty-five volts are siphoned off.

Now imagine that your thoughts begin to gently drift away from the project. Your mind wanders to other things: an upcoming vacation, your mortgage, a recent movie, or the unwashed dinner dishes. As the project in front of you fades into the background, you lose another twenty-five volts. Three quarters gone.

And imagine that from a distant corner of the house, in another room, you hear the faint rhythm of a dripping water faucet. If this continues, you might have no attention left at all!

For much of the time, we travel through life using only a portion of our mental energy. Because we tend to be busy inside our heads — planning, anticipating, analyzing, worrying — our thinking becomes cluttered. At times, it's as if we have several unrelated pools of activity inside our head. One part of our mind is thinking about the future. Another part is recalling the past. Another part is holding a conversation with still another part of our mind. Our thinking becomes cluttered with a forward rush of words and images. Like bouncing footballs, our thoughts jump quickly in unpredictable directions. We lose some of our full hundred volts of attention to distractions and preoccupations. As a result, we don't work at peak efficiency. For this reason, before exerting your mind, it's a good idea to loosen up mentally.

Loosening up is an important part of any workout — it prepares your system for exertion. To warm up before running a mile, you stretch your arms, legs, shoulders, and back. This increases circulation to your limbs, making your muscles looser and less likely to become stiff afterwards. Similarly, when you stretch your mental muscles before a mental workout, you also enhance your performance.

Not surprisingly, being loose is part of being fit. It's a feeling of well-being. When you're loose, your muscles don't work *against* each other; they work *with* each other. As a result, your movements are smoother, more natural. When you don't fight with yourself, you feel at ease, unrushed, unpressured and uninhibited.

How do you loosen up mentally? One approach is to collect your attention and bring the focus of your awareness into the here and now. You can do this by relaxing, slowing down, and allowing your thoughts, concerns, and tensions to pass right through your mind. You can also do this by fixing your attention on a single specific object or task. Whichever approach you use, the trick is to let go of the irrelevant thoughts that clutter your thinking. Do this and you will regain more of your hundred volts.

Let's begin with the letting go approach. A natural starting point for loosening up is to relax your body. By releasing the tension in your shoulders, by breathing more deeply and evenly, and by un-clenching the tiny muscles around your mouth and eyes, you release the physical expression of tension. Relaxing your body automatical-ly relaxes your mind.

"When we learn to relax
the body, breath, and mind,
the body becomes healthy,
the mind becomes clear,
and our awareness becomes balanced."
TARTHANG TULKU, *Buddhist teacher*

The Grand Tour

Close your eyes, settle into a comfortable posture,
and spend a few minutes relaxing your body.

Begin by letting your body become loose and limp.
Allow your weight to sink
and your muscles to relax.

Spend a while just paying attention to how your body feels.
Focus on your physical sensations,
in your arms, shoulders, back, head, stomach, and legs,
as well as inside your chest, abdomen, and hips.

Then slowly shift your attention to your breath.
Focus on the sensation of air
passing through your nostrils.

As you inhale and exhale, allow your breathing
to become calmer and more even.
Don't try to force your breath.
Just allow it to be natural and fluid.

Each time a distracting thought passes through your mind,
use it as a reminder
to return your attention to your body.
Gently lead the focus of your mind back to your sensations.

Allow yourself to let go completely
and to sink deeply into the warm feeling of relaxation.
Recirculate your sensations back into your sensations.

Become so quiet inside
that you can feel
your heart beat
throughout your body.

As your attention becomes clearer with each breath,
turn it to relaxing specific parts of your body.

Begin by mentally picturing your face.
Visualize your eyes, mouth, cheeks, and jaw.
Form a vivid mental image of each part
becoming more relaxed as you gaze on it.

As you turn your attention to these parts of your face,
you may discover the presence of subtle tensions.
Simply allow the tensions to dissipate
through the visualization.

When your face is thoroughly relaxed,
move on to your ears, neck,
shoulders, arms, and fingers.
Visualize each part becoming looser
and more relaxed.
The clearer your picture, the more deeply you relax.

Continue visualizing the rest of your body:
your chest, back, stomach, legs, knees, and toes.
Remember, there is no need to rush,
just let yourself enjoy
the experience of touring your body.

Once you finish picturing your toes,
visualize your entire body
as a relaxed, sentient statue.
Immerse yourself in the sensations
of full relaxation.

Just let go.

Put the book down,
take ten or fifteen minutes,
and loosen up your body,
inside and out.

Physical Relaxation

When you sit still and calm your body, you give yourself the opportunity to relax into the present moment. Your senses become more acute, your thoughts become less urgent and demanding, and your mind becomes fresher and more alert. You feel at ease and enjoy a frame of mind in which you don't have an urge to do anything else, to go anywhere else, or to have anything else. You are just there, completely mentally present.

The key to relaxing your body is to focus on the physical sensation of relaxation. If you follow what's going on in your head, you may tend to be distracted by the momentum of your thoughts. But if you sink into your sensations, paying close attention to what your body feels like — on the surface and deep inside — your thinking begins to slow down.

A good way to focus on physical sensation is to concentrate on the rhythm of your breathing. There's a subtle link between your breath and your mental state. When you're agitated, your breathing tends to be shallow and uneven, and when you're calm and collected, your breathing tends to be deep and even. What this means is that when you want to collect your attention — say, before a meeting, during a troubling encounter, or before a test — you should relax your body and steady your breath. For a few moments, stay with the sensations of your breathing. Allow it to become slow and rhythmic. Don't force it. Just allow it to be natural.

TIP Settle into the rhythm of your breathing.

Physician Rolf Alexander has an interesting relaxation technique that can be applied anywhere, any time. He suggests visualizing a double cross through your body. Picture a vertical line extending through your spine, from your tail bone to the top of your head. Picture a horizontal line passing through your chest, from one shoulder to the other, and another horizontal line through your hips, from one leg socket to the other.

Visualize that the cross is made from a strong and flexible metal that twists and flexes as your body moves. To relax, you simply allow the cross to find its natural position, your spine and head become aligned upright, your shoulders settle to the same height, and your hips become square.

If you picture the entire cross suspended from a point above your head, you have a handy way to assume a good, relaxed posture. Your head extends slightly up and out from your chest, opening up your shoulders. Your arms hang freely. Your hips keep you straight.

TIP Make it a habit to keep your body loose.

Loosening Up Inside

Though the idea of mentally letting go is simple, the action is not so easy. If you're like most people, you'll find that after a few minutes, your thoughts become distracting. Your mind starts to plan, to anticipate, or to work out problems. You may become entranced with a stream of words and images. You may begin to observe how well you are relaxing. You may want to feel a certain way and begin to instruct yourself to reach some mental or emotional state. These urges, though subtle, prevent you from letting go fully.

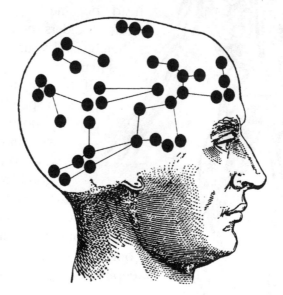

Loosening up your mind does not mean stopping your thoughts. Trying to stop your thinking completely is about as hard as trying to stop your breathing, and maybe just about as useful. Loosening up your mind involves *letting go of the urge to follow* each and every thought that passes through your brain. You let them pass freely, one at a time, but you let go of the need to act on them.

How do you free yourself from the magnetic pull of everyday thoughts? One good way is to count. With each out breath, silently sound a number in your mind. Slowly count up from one to ten, then down from ten to one. In between the numbers, allow your regular thoughts to pass, but then return your attention to the numbers. Like the rhythm of breaking surf, the rhythm of your words has the power to soothe and relax, to keep you afloat above mental tides and currents.

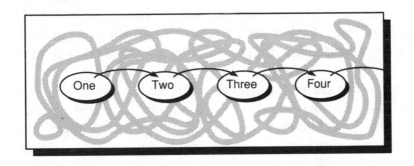

Another way to let go is to visualize your mind as a wide open blue sky, and individual thoughts as birds, first emerging from the distance, flying overhead, and then disappearing back into the distance. When a thought appears, you allow it to think itself through at its own rate. You don't try to rush it; you let it pass overhead.

When you watch your thoughts closely in this way, you'll realize that each has its own character. Some are fast, others are slow. Some refer to the future, some to the past. By learning to stay with your thoughts, without trying to manipulate, analyze or sort through them, you'll learn to experience directly their tones and patterns.

TIP Remember the words of Buddhist teacher Tarthang Tulku, "In your mind, what is happening is what you are doing."

Mental Core Dump

One way to loosen up your mind is to become aware of all the demands it perceives. On a single piece of paper, write down *everything* that's on your mind: long-term goals, short-term goals, nagging desires, things you'd rather have said, things you have to or want to do, shopping lists, house decorating ideas. No matter how apparently trivial, write down anything that you have to do or need to resolve. Jot down these items in point form — a word or two for each thought will suffice. Continue until you have nothing more to write.

Writing down everything that's on your mind has the psychological effect of removing mental clutter. When you can see, at a glance, all the things that you have to deal with, as well as the things that you have been subconsciously thinking about, you can confront them directly. With the demands out in the open, you counteract the feeling that you have forgotten something or that you are avoiding something. You can then make decisions, set priorities, and free up your attention to deal with what's at hand now. If nothing else, you can tell your mind that you will return to the demands later.

Think List

Figure out finances
Write a letter to Charlie
Talk to Chris about deadline
Take car in for oil change
Call Sue at the library
Get Danuta's present
Fix the bathroom faucet
See Maya and Nika
Get a map for the trip
Call Christopher
Work out

The Magic Massaging Finger

Visualize your brain inside your head.
Picture where it lies behind your eyes,
on top of your spine.

Now visualize a finger that has the power
to massage and soothe away any tensions.

Begin by massaging the outer layer of your brain.
Focus on the relaxing sensations of your finger.
Let the tension be eased away.
Allow the fresh energy, released by the massage,
to flow towards the center of your brain.

Allow the tangible, warm, and tingling
feelings to expand.

Massage your brain in a series of layers,
beginning from the outermost layer
and progressing inward.

If distracting thoughts interfere,
just view them as cerebral sensations.
Visualize that you can massage
any part of your brain
and massage out distracting thoughts.

Afterwards, just experience
the feeling of inner relaxation.

Just sit there
and enjoy where you are
right now.

Loosening Up Style Tips

There's a certain kind of magic that happens when we're loose. The mind becomes calmer, attention becomes clearer, and we become more ready to respond to anything that may come our way.

TIP ONE Keep your mind fresh by relaxing your body. Settle into a comfortable posture, relax any muscles you don't need to use, steady and even out your breathing, and sink into your sensations.

TIP TWO Unclutter your thinking by identifying what you need to think about. Write out a list of demands, needs, wants, pressures, anticipations, anything that's on your mind. Release the hidden concerns by looking at the big picture.

TIP THREE Bring your mind into the here and now. Develop a technique to come back to your senses. Learn the best way for you to collect the full hundred volts of your attention and to focus it where and when you want it.

TIP FOUR Periodically, take some time to relax and unwind. Develop the skill of letting go fully by setting aside some time each day to do nothing.

"The bow that's always bent will quickly break;
But if unstrung will serve you at your need.
So let the mind some relaxation take
To come back to its task with fresher heed."
PHAEDRUS, *1st-century poet*

2

MENTAL MOVEMENT

Concentration

The Two-Minute Mind

*Sit in front of a clock or a watch
that has a sweep second hand.*

*Relax for a few moments,
collect your attention,
and when you're ready,
place your attention on the motion
of the second hand.*

*For two minutes,
focus your awareness
on the movement of the second hand
as if nothing else in the universe existed.*

*If you lose the thread of concentration
by thinking about something else,
or just by spacing out,
stop,
collect your attention,
and start again.*

*Try to keep your concentration
for two solid minutes.*

*Stop reading,
get a watch or clock,
and go ahead and do it.*

Start now.

Attention:
Your Most Important Muscle

*"The essential achievement of free will is to attend
to a difficult object and hold it fast before the mind."*

WILLIAM JAMES, *19th-century psychologist*

What skill could be more essential to mental fitness than the ability to concentrate? Concentration, the ability to tune in some things and tune out others, underlies every other skill. It enables you to reason, to think clearly, to drive a car through a busy intersection, to plan your finances, to learn a new dance step, and to solve a differential equation. It lets you hear the distant call of a yellow-bellied warbler above the sounds of rustling leaves and to distinguish the fine flavors in a Bordeaux wine.

How did you do with the Two-Minute Mind? Was it easy to concentrate at first, then progressively more difficult? Did you find yourself thinking about how well or how poorly you were doing? How long a stretch could you hold before something else caught your attention?

No matter how well or how badly you think you did in your first try, you probably noticed that, after awhile, your mind wanted to move on to something else. While passive attention — the kind of attention that's involved in noticing movement, listening with half an ear — happens automatically, active attention demands deliberate exertion.

The truth of the matter is that your attention constantly shifts. It's dynamic, ever on the move, focusing on one thing one moment and on another the next. It is the nature of attention to wander, recall, and anticipate. This mental movement gives a sense of continuity, context, and perspective to your world.

In this way, attention is like eyesight. Both attention and vision are selective: at any one instant, you see details within the central portion of your entire visual field. What is at the center of your vision — like the sentence you are now reading — is clear and distinct, and what is at the periphery of your vision — like the rest of the room — is vague until, of course, you look directly at it. Similarly, what you attend to — like what you are now reading — is clear in your consciousness, and what you don't attend to — say, the weight of your clothes — is less clear.

To build up a full picture of what you see, your eyes dart around, settling here for a moment, there for another moment, painting in a complete wide-angle image. Similarly, to get a sense of context, your attention moves around, focusing on one thought at one instant, zooming in on something else the next instant. It anticipates, shifts, and moves to get a full perspective on what's going on.

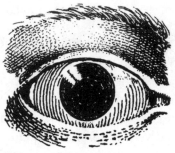

There are definite limits to our attention. We can concentrate only so long on something before our mind skips on to something else. Another limitation is that we can juggle only so many items at one time. To demonstrate this, try the following exercise. Read each of the following series of numbers to yourself. After each series, close your eyes, and repeat the numbers. How long a series can you hold in your mind?

3 5 4 8 4

5 7 9 1 3 2

2 5 4 7 7 0 4

8 5 7 1 3 2 7 0

2 4 6 5 8 4 2 4 5

1 2 6 1 9 4 1 7 2 1 1 9 6 9

If you're like most people, you may find that juggling five numbers in your head is manageable. Seven numbers becomes fairly difficult, and fourteen numbers seems to be next to impossible. Psychologists suggest that most of us can carry, at most, about seven discrete bits of information at once. We can deal fairly easily with a seven-digit telephone number, with seven countries in a continent, with seven new people in a meeting. Any more than that and we need either to write things down or to rearrange the information in a more manageable way.

One way to increase attention span is to organize information into meaningful groups. For example, if you think of the sequence of twelve digits as some important dates — with the first seven numbers, 12–6–1941, representing the bombing of Pearl Harbor, and the second six numbers, 7–21–1969, representing the first manned lunar landing, you will probably have less difficulty recalling it.

Because our attention can hold only so much, and because our attention shifts from one thing to another, we must deal with one inevitable consequence: our mind *tunes in* what we want *tuned out,* and *tunes out* what we want *tuned in.* Our attention wanders.

However, we can use this tendency to enhance our concentration. Psychologist William James summed it up this way: "There is no such thing as voluntary attention sustained for more than a few seconds at a time. What is called sustained voluntary attention is a repetition of successive efforts which brings back the topic to the mind. The topic once brought back, if a congenial one, develops: and if its development is interesting it engages the attention passively for a time." The trick is to set up a mental rhythm.

When you have a specific task to perform — it can be anything, mowing a lawn, writing a memo, painting a door frame, listening to a lecture — first decide how long and how well you want and need to concentrate. Then keep your mind focused. If you discover that your mind is beginning to wander to daydreams or distractions, gently return it to the task at hand. Make it a habit to pay complete attention to what's happening, and what you're doing, first for a short duration, and then for longer, and you'll find it increasingly easy to flex your mental muscles when you really need them.

The Two-Minute Mind is one of the best exercises to train attention. Curiously, if you begin practicing the Two-Minute Mind, you may discover that the exercise soon becomes harder. This is because with practice you become more discerning of your own attention. You know exactly when your attention is murky and when it's clear. You become more critical and demand of yourself a higher degree of mental clarity. After a couple of weeks of steady practice — say five or ten minutes a day — you'll feel a remarkable improvement in your attention. You will be able to concentrate for longer periods. Your mind will feel clearer, and it will take less time to gather and focus the full hundred volts of mental energy.

To make the basic exercise more interesting and more challenging, here are a few variations:

• Place the clock or a watch directly in front of a television playing a sit-com, the news, or better yet, commercials. Try to focus your attention on the movement of the second hand for two solid minutes. Don't allow the television to steer the focus of your attention.

• Focus half of your attention on the motion of the second hand and half on your hands. Split your attention down the middle.

• Place half of your attention on the motion of the second hand and half on a number series. Mentally recite the numbers 2, 4, 6, 8, 10, 8, 6, 4, 2, 4, 6, and so on, juggling both items in your mind. If you start thinking about something else, or if you lose your place in the series, start again. Strive for two minutes or longer.

• Concentrate on the motion of the second hand with a third of your attention. With another third, focus on reciting a verse, like Mary Had a Little Lamb, or Row, Row, Row Your Boat. With the remaining third of your attention, focus on a number series.

Physical Postures: One Thing at a Time

If you watch a diamond cutter shape a precious stone or a conductor lead a symphony orchestra, you'll realize that a person who concentrates well doesn't waste physical movements. Watch a person who fidgets, scratches, constantly shifts his weight, or holds his muscles needlessly tight, and you'll see that one sign of scattered attention is scattered movement.

When you need to concentrate, assume a physical posture that makes it easy to focus attention. Relax your muscles, picture the double cross through your shoulders and hips, and avoid restless movements. If you're working at a desk, plant your feet on the ground, sit squarely, lean slightly forward, and concentrate on what's in front of you. Do this for a few minutes and you become more focused. When you appear relaxed yet attentive, you tend to become more relaxed and attentive. With a physical posture conducive to concentration, mental muscles work more productively and efficiently.

"When sitting,
just sit.
When standing,
just stand.
Above all,
don't wobble."

ANCIENT CHINESE SAYING

Emotional Postures: Finding Incentives

The harder you push your mind, the harder it becomes to concentrate. Concentrating can be like getting a donkey to move. Push or pull and it resists. Push harder and it resists more: you end up only fighting with yourself. The trick to getting the donkey to move is not to use force, but to use encouragement. Dangle a carrot in front of its nose, and it will follow you anywhere.

Interest is the emotional component of attention. You don't have to force your mind to concentrate on a thriller novel or on an action movie. You don't have to compel your mind to focus on anything you've got a stake in. Interest powers your mental muscles. And since it is natural to pay more attention to what's important to you, the way to deepen your attention on something that you don't care a lot about is to get your mind interested. With interest, there's no problem with concentration. Your attention naturally follows.

Cultivating interest is largely a matter of adopting an outlook of curiosity and inquisitiveness. Imagine walking along a pebbly beach. With an inquisitive mindset, you notice the pebbles. You see differences and similarities. Some pebbles are large and rough, some are small and smooth. You pick up a single stone and see flecks of crystal. You find tiny hidden caverns and gleaming surfaces and discover a previously hidden world of cracks and fissures, of reflections, of broken symmetries. The more you look, the more you see. The more you see, the more you want to keep on looking.

Becoming inspired by everyday tasks — such as doing the dishes, mowing the lawn, working on another financial report — involves seeing. You lead your mind to find the interesting aspects of the task. You look for the unusual. You contrast what you know with what you don't know. And as you allow your mind to make connections, you will shift mental postures and become curious and even intrigued by something you originally found dull. If you want to keep your attention on something, try to find out something new about it.

Next time you feel bored, say in a meeting or in a conversation, slip out of your boredom by looking for the interesting aspects — even if your mind doesn't want to. In a meeting, ask yourself how the situation might look if you saw it through the eyes of a four-year-old, of a person who didn't know the people in the room, or someone who didn't speak the language. If you find yourself trapped with a less than inspiring companion, try to turn the conversation into a genuinely interesting one. Ask the person what he hates, what he enjoys, or what's important to him, and you're likely to discover that he is not as boring as you thought.

Besides having the ability to create and sustain interest, a fit mind has a full storehouse of personal interests. It provides a fertile environment for ideas to take root, germinate, and flower. William James put it this way: "Attention is easier the richer in acquisition and the fresher and more original the mind. And intellect unfurnished with material, stagnant, unoriginal, will hardly be likely to consider any subject long."

What kind of things might a fertile mind find interesting? Virtually anything under the sun can be food for thought: ancient history, relativity, bioengineering, Third World economies, cattle judging, archeology, ancient religions, rattlesnake bite cures, the making of balloons, high fashion, aerodynamics, house construction. As an exercise in fertilizing your imagination, go into a bookstore and pick up a book on a subject about which you know nothing, say gardening or lamination techniques. Or buy a magazine on graphic design, or computer software, or wilderness camping. Venture beyond your ordinary sphere of interests and feed your mind with new information.

*"The happiest person is the person
who thinks the most interesting thoughts."*
TIMOTHY DWIGHT, *19th-century educator*

Mental Postures:
Setting Priorities

Good concentration — the kind that inspires efficient and productive work — is like good time management. Both depend on knowing what needs to be accomplished, on establishing priorities, and on setting goals. When you have a clear idea of what's important, you can economize your resources and prevent them from slipping away.

Before work, it's a good idea to pause and allow your thinking to become smooth and unhurried. Mentally stand back, look at your agenda, and rate the importance of each item on your list. Ask yourself whether something is:

If something is indeed important, then it's worth paying attention to. If something is not so important, then you can take care of it once you have the important things out of the way. If something is trivial, leave it for the end of the day.

Once you've established your priorities, give yourself specific objectives. Plan out what you wish to accomplish by setting definite tasks. Form a mental image of the completed objective. With that image, follow the task through, one step at a time, until it's finished. Break down large tasks into smaller sub-tasks. Encourage your mind to work by concentrating on details. With bite-size tasks, your mind will tend to wander less. You may become surprised at how efficient you can be when you economize your mental energy.

To keep your mind focused on what you've got to get done, do your best to remove any distractions. If you can, close your door, have your calls screened, and avoid the urge to get up and talk to someone. It's helpful to arrange your time so that all the trivial stuff you have to get done — phone calls, cleaning, and so on — is lumped together in one block of time.

Determine your personal rhythm for working. Do you get more creative ideas in the morning, the afternoon, or late at night? Do you work best at organizing little details in the beginning of the day or at the end? Schedule your time to take advantage of your personal rhythms.

TIP To focus your mental energy better, remember to adopt a conducive physical posture, to generate interest, and to give yourself specific tasks. Armed with these techniques, your mind will become sharper.

"Concentration is the secret of strength
in politics, in war, in trade,
in short, in all management of human affairs."
RALPH WALDO EMERSON, *19th-century philosopher*

41

Attention Grabbers

Here are some more exercises to help you practice focusing your attention. Try them when you're feeling up and on the ball, then try them when you're feeling distracted and out of sorts.

❑ **Action Cycles**
Next time you're washing dishes, economize your attention by dividing the routine into cycles of action. When you take a spoon to be washed, mentally sound the word "start." Wash the spoon with the quality of attention you'd use if you were performing brain surgery. When you're done with it, put the spoon in the drying rack, and mentally sound the word "stop." Then get the next object, and repeat the process.

❑ **Mental Coffee Breaks**
Place a small object, like a pen, coin, or paper clip on a table top in front of you. For five minutes, narrow your attention onto the object. Each time your mind skips onto something else, gently lead it back to the object. Count the number of times your mind jumps.

❑ **Mindbeats**
Slowly draw a pencil across a blank piece of paper. Concentrate on keeping your attention on the place where the pencil point becomes a line. Each time your mind wanders off, draw a mindbeat: mark the place with a kink in the line. When you reach the edge of the paper, double back. How long can you keep an unbroken line of awareness?

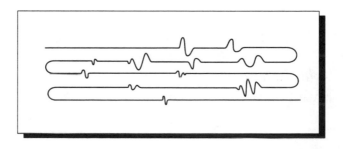

❏ **Center of the Universe**
Next time you're riding in a bus or subway and have some time to kill, look around you and select an object, like an advertisement, the back of someone's head, some marks in the ceiling. For five minutes, focus on this object to the exclusion of everything else. Let there be nothing else in the universe. Even though your mind will want to move to other things, concentrate only on the object you've chosen. Only when your time is up, relax and look around.

❏ **Sasaki's Saying**
Photographer Chris Sasaki once said that whenever he caught himself woolgathering — immersed in a frame of mind that's accompanied by a vacant stare and a dim awareness of the passing of time — he would sound the word "attention" clearly in the middle of his head. Then he would look around and really notice where he was and what he was doing. He found this exercise a good way to become more alert.

❏ **Metattention**
Next time you browse through a magazine or newspaper, notice what you notice. Notice what ads, articles, and pages your eye is drawn towards. What does it feel like to have your attention diverted to something? What part of your mind is controlling your attention?

❏ **In the Timing**
Choose something in your immediate field of view, say a pencil, and in your most persuasive voice, say to yourself, "Look at that pencil." Then pause for a moment, and wait for your attention to respond. The pencil will appear brighter in your awareness. Then acknowledge this response by mentally saying "O.K." Then repeat it a few more times. "Look at that pencil." Pause. "O.K." "Look at that pencil." Pause. "O.K." "Look at that pencil." Pause. "O.K." Pay attention to the quality of your attention. Notice the response time, how long it takes for your mind to respond. Does your mind respond faster with practice? Look around you now and focus your attention on something.

☐ Counting on Distractions

If you're reading a difficult text and you're finding that your mind is wandering all over the universe, try this trick. Place a check mark in the margin of the book at the place where you noticed that your mind drifted. Go back to where you can remember reading, and continue from there. When you reach the bottom of a page, mentally review what you've just read. If you can't recall the main ideas, return to the top of the page and reread it. If you persist, you'll probably find that your level of comprehension gets higher and the check marks become fewer.

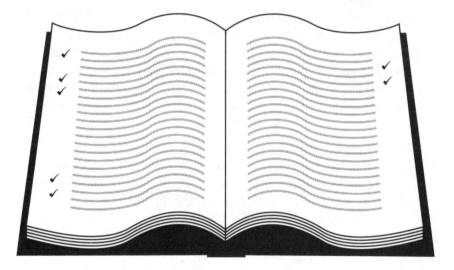

☐ Creative Tension

If you find yourself daydreaming when you don't want to, intentionally put your body in a posture that you don't ordinarily assume. Cross your legs the other way, shift your back slightly, adjust the position of your feet, or put a slightly different expression on your face. If you happen to be in a room with other people, discreetly mimic another person's posture. When you put your body into an unfamiliar position, it tends to want to fall asleep less. If you need extra energy, then tighten the muscles in your stomach, in your buttocks, or in your legs. Make it hard for your body to become drowsy, and you make it easy for your mind to become alert.

Noticing More of What's Happening

EXERCISE Look around the room and find six objects that contain circles.

O O O O O O

With a circle mindset, you'll find circles jumping out all around you — the top of a cup, the end of a pencil, the face of a screw in a light switch. When you set your attention to look for something, you almost always find it. You just need to know what you're looking for.

What you pay attention to determines what your world is like. Focus on problems, and your world is full of obstacles. Focus on nothing in particular, and your world is a clutter of unrelated experiences. Focus on creative ideas, and your world opens to a realm of unlimited possibilities. People find what they are looking for.

What do you pay attention to all day long? If you could somehow record on a video tape what you notice, what would the tape show? What would it miss? If you played back a tape of what you notice on your daily trip to work or school, would there be a continuous stream of images, or would there be gaps in the movie? Would only a few key landmarks stand out in an ocean of grey? How much detail would show in people's faces? Would the sound be muffled, or would it be in stereo?

To become more aware of what's happening around you, periodically stop what you're doing, look, listen, and ask yourself, "What's happening?"

TIP Take a closer look at the familiar.

45

Concentration Style Tips

Pay attention to your attention — it is your most important mental muscle. Encourage it to become stronger and more flexible by periodically concentrating with all of your energy.

TIP ONE Adopt a physical posture that makes it easy to concentrate. Relax your body and avoid wasted movements. Pablo Picasso said, "While I work, I leave my body outside the door, the way Moslems take off their shoes before entering the mosque."

TIP TWO Adopt an emotional posture that makes it easy to concentrate. Become interested in the task at hand. Gently lead your mind through encouragement, not through force. Look for the interesting aspects of the task and relate them to your other interests.

TIP THREE Adopt a mental posture that makes it easy to concentrate. Identify your tasks clearly, and set definite objectives. Find a mental rhythm that builds a momentum to your work.

TIP FOUR When you can't seem to concentrate, fake it. Pretend that you are absorbed in what you're doing. Ask yourself what position would your body be in if your mind was focused? What would it feel like to be interested? What would you think about if you were economizing your attention? If you continue this, an attitude of concentration will naturally evolve.

> *"If there's one thing worth hoarding,*
> *it's your attention."*
> BILL HARVEY, *author*

3

MENTAL CALISTHENICS

Increasing Your Endurance

Mental Staying Power

In your head,
count the number of capital letters
of the alphabet
that contain curved lines.

Start now.

Building Up Mental Power

*"For purposes of action nothing is more useful
than narrowness of thought combined with energy of will."*

HENRI FREDERIC AMIEL, *19th-century philosopher*

In the second half of the seventeenth century, Isaac Newton, the father of classical physics, was given a mathematical problem by the Royal Society. The problem, which had been tackled by the best minds in Britain for a number of months, was to find the formula that described the shape of the curve of a hanging string. Newton received the problem in the morning and sat at his bedside, completely still and blotting out the rest of the universe, until he'd found the answer. By dinner time, Newton had not only solved the problem, he had also developed and applied differential calculus.

Newton had many traits that distinguished him as a mental giant. He wrote and thought deeply about philosophy, optics, physics, as well as mathematics. It's said that he found it strange that Euclid, the third-century B.C. mathematician, had bothered to write out the *Elements*, his classic work in geometry, when its theorems all followed naturally from five axioms. But perhaps Newton's greatest gift was his remarkable ability to concentrate for long periods. In his own words, "If I have done the public any service, it is due to patient thought." In short, Newton's mind was conditioned.

All kinds of action require mental conditioning: getting columns of numbers to balance, working through the options in a commodities market, wrestling down four term papers and six exams in two weeks, any kind of thinking that demands an increased level of activity for a long stretch of time.

There's no short cut for conditioning. A conditioned body, with a strong heart and open lungs, reaches that state through aerobic exercise: running, swimming, jogging, rowing, cycling. A conditioned heart beats easily during normal activity, but responds quickly to increased demand. Fit lungs deliver a greater amount of oxygen to the blood. Fit cells can assimilate food faster. To get your body in this shape, you need to exert it for a sustained period. Exertion is the cornerstone of physical conditioning.

Similarly, a conditioned mind grows fit by regular exertion. You must put yourself in the position where you need to concentrate for a sustained time. You must persist. You must guide your mind to go where you want it to go. With practice, your mind becomes used to concentrating, and your thinking becomes clearer.

Calculation is a tried and tested way to add backbone to your thinking and to condition your mind. Plato understood the value of exercising calculation muscles when he wrote in *The Republic*, "Those who are by nature good at calculation are, as one might say, naturally sharp in every other study, and those who are slow at it, if they are educated and exercised in this study, nevertheless improve and become sharper than they were."

Speaking of calculation, how did you do with the letter puzzle? There are in fact eleven capital letters in the alphabet that contain curved lines:

B C D G J O P Q R S U

Many of the following exercises are adapted from the work of A.R. Orage, author of *Mental Exercises and Essays*, published in 1930. Each of the four kinds of exercises — those with numbers, letters, words, and verses — demands that you flex your calculation muscles. Many of the exercises are designed so that if you happen to lose the thread of your attention, you lose your place in the calculation. You should know exactly where your mind left off, so you can return and try to sustain your concentration for longer. As your calculating muscles become stronger, try the more difficult exercises.

These drills are to mental exercise what jogging is to physical exercise. They encourage persistence. You can perform the exercises silently or aloud, fast or slow. This makes them ideal for long rides on a bus or subway. You may very well be surprised just how few days it takes to progress to exercises you thought you'd never be able to do.

Number Exercises

☐ **Recite the following number series:**
up by 1: 1, 2, 3, 4, ..., 100
down by 1: 100, 99, 98, 97, ..., 1

☐ **Recite the ascending and descending series:**

up by 2: 2, 4, 6, 8, ..., 100   *down by 2:* 100, 98, 96, 94, ..., 2
up by 3: 3, 6, 9, 12, ..., 99   *down by 3:* 99, 96, 93, 90, ..., 3
up by 4: 4, 8, 12, 16, ..., 100   *down by 4:* 100, 96, 92, 88, ..., 4
up by 5: 5, 10, 15, 20, ..., 100   *down by 5:* 100, 95, 90, 85, ..., 5
up by 6: 6, 12, 18, 24, ..., 96   *down by 6:* 96, 90, 84, 78, ..., 6
up by 7: 7, 14, 21, 28, ..., 98   *down by 7:* 98, 91, 84, 77, ..., 7
up by 8: 8, 16, 24, 32, ..., 96   *down by 8:* 96, 88, 80, 72, ..., 8
up by 9: 9, 18, 27, 36, ..., 99   *down by 9:* 99, 90, 81, 72, ..., 9

☐ **Recite the double ascending series:**
up by 2, 3: 2-3, 4-6, 6-9, 8-12, ..., 66-99
down by 2, 3: 66-99, 64-96, 62-93, 60-90, ..., 2-3
up by 3, 2: 3-2, 6-4, 9-6, 12-8, ..., 99-66
down by 2, 3: 99-66, 96-64, 93-62, 90-60, ..., 3-2
up by 3, 4: 3-4, 6-8, 9-12, 12-16, ..., 75-100
down by 3, 4: 75-100, 72-96, 69-92, 66-88, ..., 3-4
up by 3, 5: 3-5, 6-10, 9-15, 12-20, ..., 60-100
down by 3, 5: 60-100, 57-95, 54-90, 51-85, ..., 3-5
up by 8, 3: 8-3, 16-6, 24-9, 32-12, ..., 96-36
down by 8, 3: 96-36, 88-33, 80-30, 72-27, ..., 8-3

☐ **Recite the double alternating series:**
up by 2, down by 2: 2-100, 4-98, 6-96, 8-94, ..., 100-2
up by 2, down by 3: 2-99, 4-96, 6-93, 8-90, ..., 66-3
up by 3, down by 4: 3-100, 6-96, 9-92, 12-88, ..., 75-4
up by 3, down by 5: 3-100, 6-95, 9-90, 12-85, ..., 60-5
down by 5, up by 4: 100-4, 95-8, 90-12, 85-16, ..., 5-80
down by 7, up by 6: 98-6, 91-12, 84-18, 77-24, ..., 7-86

Recite in ascending order the numbers from 1 to 100:
containing the digit 7 (or 5 or 9 or 1)
containing either the digits 4 or 6 (2 or 5, 6 or 1)
the sum of whose digits is 7 (or 6 or 9 or 8)
the sum of whose digits is divisible by 3 or 4

Recite a triple ascending series:
up by 2, 3, 4: 2-3-4, 4-6-8, 6-9-12, 8-12-16, ..., 48-72-96
up by 2, 3, 5: 2-3-5, 4-6-10, 6-9-15, 8-12-20, ..., 40-60-100
up by 3, 8, 7: 3-8-7, 6-16-14, 9-24-21, 12-32-28, ..., 36-96-84
up by 9, 5, 3: 9-5-3, 18-10-6, 27-15-9, 36-20-12, ..., 99-55-33

Recite a triple descending series:
down by 2, 4, 3: 100-100-99, 98-96-96, 96-92-93, 94-88-90, ..., 52-4-27
down by 5, 2, 3: 100-100-100, 95-98-97, 90-96-94, 85-94-91, ..., 5-62-43
down by 7, 5, 4: 98-100-100, 91-95-96, 84-90-92, 77-85-88, ..., 7-35-48
down by 3, 2, 3: 100-100-99, 97-98-96, 94-96-93, 91-94-90, ..., 1-34-0

Recite a triple alternating series:
down by 2, up by 4, down by 3: 100-4-99, 98-8-96, 96-12-93, ..., 52-100-28
up by 3, down by 3, up by 4: 3-99-4, 6-96-8, 9-93-12, ..., 75-28-100
up by 5, down by 3, up by 2: 4-100-3, 9-97-5, 14-94-7, ..., 99-43-51
down by 4, up by 3, down by 7: 99-2-100, 95-5-93, 91-8-86, ..., 43-54-2

Recite a quadruple series:
up by 2, 3, 4, 5: 2-3-4-5, 4-6-8-10, 6-9-12-15, ..., 40-60-80-100
up by 2, 5, 4, 1: 2-5-4-1, 4-10-8-2, 6-15-12-3, ..., 40-100-80-20
up by 3, 5, 2, 7: 3-5-2-7, 6-10-4-14,- 9-15-6-21, ..., 21-70-28-98
down by 2, 3, 4, 5: 100-100-100-100, 98-97-96-95, ..., 60-40-20-0
down by 2, 3, 4, 7: 100-99-100-98, 98-96-96-91, ..., 72-57-44-0
down by 3, 5, 4, 2: 100-100-100-100, 97-95-96-98, ..., 40-0-20-60

Recite a quadruple alternating series:
up by 2, down by 2, up by 3, down by 3: 2-100-3-99, 4-98-6-96, 6-96-9-93, ...
up by 2, down by 3, up by 4, down by 5: 2-99-4-100, 4-96-8-95, 6-93-12-90, ...
down by 2, 3, up by 7, 3: 100-100-1-1, 98-97-8-4, 96-94-15-7, 94-91-22-10, ...
up by 4, 3, down by 6, 4: 4-3-98-100, 8-6-92-96, 12-9-86-92, 16-12-80-88, ...

Variations on Number Exercises

❏ **Visualize a regular series of numbers:**
Mentally picture a number sequence. Instead of reciting the numbers aloud or silently in your mind, try to *see* the shapes of the digits as if they were in front of you. To make it tougher, visualize two, three, or even four series simultaneously.

❏ **Recite one number series while writing down another:**
While reciting the ascending series, 3, 6, 9, 12, ..., write down the ascending series 4, 8, 12, 16, ..., Recite the sequence, 4, 8, 12, 16, ..., while writing down the sequence, 100, 98, 96, 94, ..., Recite the double series 2-3, 4-6, 6-9, ..., as you write down the double series, 3-5, 6-10, 9-15, 12-20, ..., Make the routine as challenging as you want.

❏ **Recite one number series while visualizing another:**
Recite the sequence 3, 6, 9, 12, ..., and form a mental picture of the series 5, 10, 15, 20, ...

❏ **Recite a series, but name only the sum of the digits:**
Example: 7=7, 14=1+4=5, 21=2+1=3, 28=2+8=10=1+0=1
up from 2 by 2 to 100: 2, 4, 6, 8, ..., 1
up from 4 by 4 to 98: 4, 8, 3, 7, ..., 8
down from 100 by 3: 1, 7, 4, 1, ..., 1
down from 100 by 6: 1, 4, 7, 1, ..., 4
alternating up by 2 and 5: 2, 7, 9, 5, ..., 8
alternating up by 4 and 7: 4, 2, 6, 4, ..., 9
alternating down by 2 and 3 from 100: 1, 8, 5, 3, ..., 0

❏ **Recite all the numbers from 1 to 100:**
at numbers divisible by 3 — raise your left hand
at numbers divisible by 4 — raise your right hand
at numbers divisible by 3 and 4 — clap your hands
at numbers divisible by 5 — stamp your foot

❏ **Recite a number series in a different base:**
up by 3 in base 8: 3, 6, 11, 14, ...
up by 4 in base 9: 4, 8, 13, 17, ...
up by 3 in base 5: 3, 11, 14, 22, ...
up by 5 in base 12: 5, 10, 13, 18, ...

❏ **Double a number and see how high you can get:**
Example: 2, 4, 8, 16, 32, 64, ...
Example: 3, 6, 12, 24, 48, 96, ...
Example: 7, 14, 28, 56, 112, 224, ...

❏ **Find the last number in the following bounded alternating series:**
up from 2 to 100 by 2, then down from 100 to 1 by 3, then up from 1 to 97 by 4, then down from 97 to 2 by 5, and so on. What is the last number in the sequence?

❏ **Visualize a succession of scenes while reciting a series of numbers:**
Going to an art gallery
Your daily journey to work or school
Dinner at an Italian restaurant
A baseball game
An opera
Shoveling a walk free of snow

❏ **Prestidigitation**
There is a popular number puzzle in which you must use all the digits from one to nine and any combination of plus signs and minus signs to add up to 100. The digits must also remain in their original sequence. One possible solution is this: 12 + 3 - 4 + 5 + 67 + 8 + 9 = 100. This solution contains six signs, five plus signs and one minus sign. Can you find another solution that uses only three signs?

❏ **Reverse Prestidigitation**
In this problem, use the numbers from nine to one to equal 100. One possible solution is: 98 + 7 - 6 + 5 - 4 + 3 - 2 - 1 = 100. Can you find a solution that uses only four signs?

Exercises with Letters

❏ **Recite the following letter-number pairs:**
A1 B2 C3 D4 E5 ... Z26
1A 2B 3C 4D 5E ... 26Z
Z26 Y25 X24 W23 ... A1
26Z 25Y 24X 23W ... 1A

❏ **Recite the letter-number combinations for the following phrases, sentences, and quotes, by substituting a letter's number for its place in the alphabet. For example, the word "Abracadabra" would read 1-2-18-1-3-1-4-1-2-18-1.**
Niagara, roar again!
Satan, oscillate my metallic sonatas.
The five boxing wizards jump quickly.

It was an exquisite deep blue just then, with filmy white clouds drawn up over it like gauze. — *Sarah Grand*

A born coward, Darius eventually found great happiness in judicially kicking loud-mouthed nepotists openly picking quarrels, rightly saying that unkindness vitiated warring Xerxes' youthful zeal. — *Tony Augarde*

How to teach rigor while preserving imagination is an unsolved challenge to education. — *R.W. Gerard*

Truth is the shattered mirror strewn in myriad bits: while each believes his little bit the whole to own. — *Sir Richard Burton*

❏ **Mentally translate the following letters into meaningful phrases:**
12-9-22-5 14-15-20 15-14 5-22-9-12.
20-8-15-21-7-8-20 9-19 20-8-5 19-5-5-4 15-6 1-3-20-9-15-14. — 5-13-5-18-19-15-14.

6-9-18-19-20 20-8-15-21-7-8-20-19 1-18-5 14-15-20 1-12-23-1-25-19 20-8-5 2-5-19-20. — 1-12-6-9-5-18-9.

20-8-9-14-11-9-14-7 9-19 12-9-11-5 12-15-22-9-14-7 1-14-4 4-25-9-14-7 — 5-1-3-8 15-6 21-19 13-21-19-20 4-15 9-20 6-15-18 8-9-13-19-5-12-6. — 10-15-19-9-1-8 10-15-25-3-5.

56

□ **Read the following verse using the letter-number association:**

The Siege of Belgrade

An Austrian army awfully array'd
Boldly by battery besieg'd Belgrade;
Cossack commanders cannonading come.
Dealing destruction's devastating doom.
Every endeavor engineers essay —
For fame, for fortune fighting — furious fray!
Generals 'gainst generals grapple — gracious God!
Infuriate — indiscriminate in ill,
Kinsmen kill kindred, kindred kinsmen kill.
Labor low levels longest, loftiest lines —
Men march 'mid mound, 'mid moles, 'mid murd'rous mines.
Now noisy noxious numbers notice naught,
Of outward obstacles opposing ought;
Poor patriot! partly purchas'd, partly press'd,
Quite quaking quickly, "quarter, quarter," quest.
Reason returns, religious right redounds,
Suvarov stops such sanguinary sounds.
Truce to thee, Turkey, triumph to thy train,
Unjust, unwise, unmerciful Ukraine,
Vanish vain vict'ry, vanish vict'ry vain.
Why wish we warfare? wherefore welcome were
Xerxes, Ximenes, Xanthus, Xaviere?
Yield, yield, ye youths, ye yeomen yield your yell;
Zeno's, Zorpater's, Zoroaster's zeal
Attracting all, arms against acts appeal.
— attributed to Alaric Watts, 1817

□ **Recite the following letter series:**
The alphabet alternating forwards and backwards: A, Z, B, Y, C, X, ...
As above but two letters at a time: A-B, Z-Y, C-D, X-W, E-F, U-T, ...
Alternating two forward series: A-N, B-O, C-P, D-Q, ...
A cyclic series: A-B-C-D-E, B-C-D-E-A, C-D-E-A-B, D-E-A-B-C, ...
Repeat other cycles such as T-O-P-S; Z-I-P-P-E-R; B-I-N-G-O;
C-O-N-S-C-I-O-U-S-N-E-S-S; A-B-R-A-C-A-D-A-B-R-A.
Repeat cyclic series backwards: A-B-C-D-E, E-A-B-C-D, D-E-A-B-C, ...
Repeat other cycles backwards such as I-R-O-N; F-L-A-S-H; C-L-O-C-K.

Exercises with Words

□ **Repeat these sentences backwards, word by word, after reading them once forwards:**

We hold the power and bear the responsibility. — *Abraham Lincoln*

I give myself, sometimes, admirable advice, but I am incapable of taking it. — *Mary Wortley Montagu*

All for one, one for all. — *The Four Musketeers*

He who desires, but acts not breeds pestilence. — *William Blake*

□ **Repeat these sentences forwards, word by word, after reading them once backwards:**

doing is answer shortest The — English proverb

yourself to yourself give but others to yourself Lend — *de Montaigne*

mouth his in fish a with up come will he and sea the into man lucky a Throw. — Arabic proverb

tailor the to elegance leave, truth the describe to out are you If — *Albert Einstein*

light own his in him show to is shadow his with person a confront To — *Carl Jung*

□ **Read correctly at sight the following passage in which each word is spelled backwards:**

ynaM a emit I evah detnaw ot pots gniklat dna dnif tuo tahw I yllaer deveileb. — *nnamppiL retlaW*

erehT era owt sdnik fo hturt, llams hturt dna taerg hturt. ouY nac ezingocer llams hturt esuaceb s'ti etisoppo si a dooheslaf. ehT etisoppo fo a taerg hturt si rehtona hturt. — *sleiN rhoB*

ytiralC, thgisni ro gnidnatsrednu era ylno elbissop nehw thguoht si ni ec-nayeba, nehw eht dnim si llits. nehT ylno nac uoy ees yrev ylraelc, neht ouy nac yas ouy evah tcerid noitpecrep, esuaceb rouy dnim si on regnol desufnoc. oT eb raelc, eht dnim tsum eb yletelpmoc teiuq, yletelpmoc llits, neht ereht si laer gnidnatsrednu dna eroferet taht gnidnatsrednu si noit-ca. tI t'nsi eht rehto yaw dnuora. — *J itrumanhsirK*

❏ **Read correctly the following unspaced passage:**

Itseemsthentobeoneoftheparadoxesofcreativitythatinordertothinkoriginally
wemustfamiliarizeourselveswiththeideasofothers. — *George Kneller*

Habitistheapproximationoftheanimalsystemtotheorganic.Itisaconfession
offailureinthehighestfunctionofbeing,whichinvolvesaperpetualself-determina-
tion,infullviewofallexistingcircumstances. — *Oliver Wendell Holmes*

❏ **Spell each of the following words backwards as you read them:**

classification	hypersonic	occipital
jealousy	marinade	claustrophobia
fineness	congenital	irretrievable
bureaucratic	forgetfulness	consternation
indiscriminate	prolongation	cogitation

❏ **Read correctly the following unspaced passage in which each word is spelled backwards:**

llAehts'dlrowaegats.dnAllaehtnemdnanemowyleremsreyalp:yehtevahrieht
stixednariehtsecnartne;dnaenonamnisihemitsyalpynamstrap.
— *mailliWeraepsekahS*

ehTsomiksedahytfifowtsemanrofwonsesuacebtisawtnatropmi:erehtthguoot
ebsaynamrofevol. — *teragraMdoowtA*

❏ **Read correctly the upside down passage:**

all about one thing. Universality is the best. *Pascal*

It is far better to know something about everything than to know

❏ **Read correctly the mirrored passage:**

I roamed the countryside searching for answers to things I did not understand
Why shells existed on the tops of mountains along the imprints of coral and
plants and seaweed usually found in the sea. Why the thunder lasts a longer time
than that which causes it and why immediately on its creation the lightning
becomes visible to the eye while thunder requires time to travel. How the
various circles of water form around the spot which has been struck by a stone
and why a bird sustains itself in the air. These questions and other strange
phenomena engaged my thought throughout my life.
Leonardo da Vinci

More Word Exercises

❏ **Doublets**

The doublets puzzle was invented by Lewis Carroll. This is how he describes it: "The rules of the puzzle are simple enough. Two words are proposed of the same length; and the puzzle consists in linking these together by interposing other words, each of which shall differ from the next word in one letter only. That is to say, one letter may be changed in one of the given words, then one letter in the word so obtained, and so on, till we arrive at the other given word. The letters must not be interchanged among themselves, but each must keep its own place. As an example the word 'head' may be changed into 'tail' by interposing the words 'heal, teal, tell, tall.' I call the two given words 'a doublet,' the interposed words 'links,' and the entire series 'a chain.' It is, perhaps, needless to state that it is *de rigueur* that the links should be English words, such as might be used in good society." Try your hand at the following doublets.

❏ **3 links**

Change OAT to RYE
Close an EYE LID
Make HOT TEA
It's HALF TIME
Gaze at the MOON BEAM

4 links

Drive PIG into a STY
REST on a SOFA
Change a FISH into a BIRD
Fasten your LIFE BELT
Drink some PORT WINE

❏ **5 links**

Go from FAST to SLOW
Take the PAWN with the KING
Transform POOR into RICH
Bake some FLOUR into BREAD
Take the FOOT PATH

Touch your NOSE with your CHIN
PITCH your TENTS
Go to the PLAY ROOM
Get COAL from a MINE
Grasp a PEAR from the TREE

❏ **6 links**

Make WHEAT into BREAD
Raise FOUR into FIVE

Show that PITY is GOOD
How BLACK is like WHITE

❏ 7 Links

STEAL some COINS
Get WOOD from a TREE
Convert an ELM into an OAK

Pit ARMY against NAVY
BUY an ASS
Show that ONE equals TWO

❏ Anagrams

Reshuffle the letters of the word ASPIRANT and you can find the word PAR-TISAN. How long will it take you to find three new words from the following list?

TREASON	WANDER	BESTIARY
MEDUSA	MASCULINE	DICTIONARY
DIRECTOR	CATECHISM	LEOPARD
CONSIDERATE	EXCITATION	LEGISLATOR

❏ Appended Anagrams

Add the letter C to each of the words below, then rearrange the letters to find a new word. Add another C, resort the letters, and find yet another word. For example, add a C to OIL and create COIL. Add a C to COIL and create COLIC.

HAT	LOUT	HATE
SEAR	TAPE	ARK
SPITE	HEAD	HERE
RILE	IRK	NOSE
LEAN	NEAR	NEAT

❏ Alphabet Advance

If you shift each of the letters of the word COLD three positions forward, so that the letter C becomes F, O becomes R, L becomes O, and D becomes G, you arrive at a new word, FROG. In the following list of words, some must be advanced, and others reversed. Can you figure out the other words?

TIGER	JOLLY	ADDER
FILLS	SORRY	FREUD
CHAIN	FERNS	

Exercises with a Verse

☐ **Memorize the following verse:**
Mary had a little lamb.
Its fleece was as white as snow.
Everywhere that Mary went,
The lamb was sure to go.

☐ **Recite the verse numbering each word:**
1 Mary 2 had 3 a 4 little 5 lamb.
6 Its 7 fleece 8 was 9 as 10 white 11 as 12 snow.
13 Everywhere 14 that 15 Mary 16 went,
17 The 18 lamb 19 was 20 sure 21 to 22 go.

☐ **Recite the verse with the number of letters in each word:**
4 Mary 3 had 1 a 6 little 4 lamb.
3 Its 6 fleece 3 was 2 as 5 white 2 as 4 snow.
10 Everywhere 4 that 4 Mary 4 went,
3 The 4 lamb 3 was 4 sure 2 to 2 go.

☐ **Recite the verse by spelling the words backwards:**
yram dah a elttil bmal
sti eceelf saw sa etihw sa wons
erehwyreve taht yram tnew
eht bmal saw erus ot og.

☐ **Recite the verse beginning with the last line:**
The lamb was sure to go.
Mary had a little lamb.
Its fleece was as white as snow.
Everywhere that Mary went,

☐ **Recite the verse line by line backwards:**
lamb little a had Mary
snow as white as was fleece its

went Mary that everywhere
go to sure was lamb the

 Recite the verse omitting every second word (third, fourth):
Mary a lamb
Fleece as as
Everywhere Mary
The was to

 Recite the number - letter pairings for the verse:
13-1-18-25 8-1-4 1 12-9-20-20-12-5 12-1-13-2
9-20-19 6-12-5-5-3-5 23-1-19 1-19 23-8-9-20-5- 1-19 19-14-15-23
5-22-5-18-25-23-8-5-18-5 20-8-1-20 13-1-18-25 23-5-14-20
20-8-5- 12-1-13-2 23-1-19 19-21-18-5 20-15 7-15

 Visualize the letters in the verse:
Don't sound the words or the letters in your mind; just see a stream of letters flowing by.

 Here are some other verses to stretch your endurance:
Jack and Jill went up the hill
To fetch a pail of water.
Jack fell down and broke his crown,
And Jill came tumbling after.

Humpty Dumpty sat on the wall
Humpty Dumpty had a big fall
All the king's horses and all the king's men
Couldn't put Humpty together again.

To be, or not to be: that is the question:
Whether 'tis nobler in the mind to suffer
The slings and arrows of outrageous fortune,
Or to take arms against a sea of troubles,
And by opposing end them?
— *William Shakespeare*

Endurance Style Tips

Mental endurance, the ability to keep thinking when your mind wants to hit the showers, is enhanced only by practice. Make it a habit to keep your mind focused on the tasks you set for yourself.

TIP ONE Regularly practice mental aerobics. Each day, do one thing that demands intense concentration.

TIP TWO Pay attention to which endurance exercises you find easy and which you find difficult. For the best results, work on those that don't come natually to you. Remember the adage, "No pain, no gain."

TIP THREE Make up your own mental conditioning exercises. Invent variations of number exercises, letter exercises, word exercises, and verse exercises. How much can you make your mind do?

"Education does not consist merely
in adorning the memory
and enlightening the understanding.
Its main business
should be to direct the will."

JOSEPH JOUBERT, *18th-century essayist*

S T A T I O N

4

MENTAL GYMNASTICS (I)

Thinking with Images

Mental Manipulation

Look at the palm of your left hand.
Study it very closely.
Notice the lines, cracks, and moles,
the patterns on your skin,
and how the play of light and shadow
creates subtle differences in color.

After a minute or so,
when you have a good idea
of what your palm looks like,
close your eyes
and try to see it in your mind.
Form a crisp mental image of your palm.

After a while,
open your eyes, look at your palm,
and compare it with your mental picture.

Notice what was present
and what was missing from your image.

Then close your eyes, and repicture your palm.
Imagine what it would look like
if you were looking at it with open eyes.

And after another half-minute,
open your eyes, and look at your palm.
Repeat this cycle about half a dozen times.
With each visualization,
try to add more clarity to your image.

The World of Imagination

"The Soul never thinks without a mental picture."

ARISTOTLE, *4th-century B.C. philosopher*

Imagery plays an important role in the mental lives of many people. Chemists use imagery to picture how molecules are linked together. Fashion designers use imagery to envision how a dress will hang. Theoretical physicists use imagery to model the abstract world of subatomic particles. Businessmen use imagery to chart the growth of markets. Chess masters use imagery to plan strategy. Movers use imagery to plan how to steer a grand piano up a winding staircase.

What exactly is a mental image? The word image comes from the Latin root word *imitari* which means to imitate. So a mental image is an inner imitation or representation of an object of sense. Within the private world of your imagination, the place where you experience imagery, you can hear, smell, taste, and feel, as well as see things that are not physically present. In your mind, you can picture a rose without actually holding one in your hand.

Besides portraying physical objects, mental images can also represent abstract ideas — the sense of freedom, the concept of power, the image of beauty can be viewed in your imagination.

How did you do with the visualization exercise? Was the image of your hand clear and steady, or blurry and unstable? Was the picture three-dimensional, like the real view, or a flat representation, like a television picture?

Perhaps no other mental skill varies so greatly among people as the ability to visualize. While some individuals claim to see clear three-dimensional pictures in their mind's eye, other people can form no image at all. Some people think almost exclusively with mental images, and others insist that the mind's eye is just a figure of speech.

No matter how well (or how poorly) you feel you can visualize, stretching and flexing your imagination muscles is an effective way to enhance your mental creativity and versatility, and to add immediacy to your thoughts.

Cognitive psychologists usually speak of the ability to visualize in terms of two factors, *vividness* and *controllability*. Vividness — how bright, clear, and lively an image appears — and controllability — how steadily and responsively an image behaves — are interrelated attributes. When you add vividness, your images are more colorful, realistic, and three-dimensional. When you achieve controllability, your images are steadier and more precise. Let's take a closer look at each of these qualities.

*"The debt we owe
to the play of imagination
is incalculable."*

CARL JUNG, *psychologist*

Vividness and Controllability

Nicolas Tesla, the inventor of the fluorescent light bulb, the AC generator, and high-voltage electricity, had an extraordinary capacity to create vivid imagery. He routinely envisioned three-dimensional images of complicated machines. These were complete in every detail and as clear as blueprints. Even more astonishingly, Tesla tested his machines mentally by letting them run in his mind for several weeks and thoroughly examining their components for signs of wear.

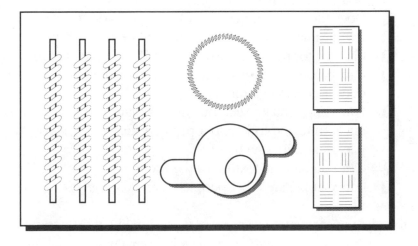

Tesla was probably an "eidetic" imager; his mental pictures were as bright and clear as his eyesight. An eidetic imager, a person who has a truly photographic memory, can look at the front page of a newspaper for a few seconds, turn away, and then read the entire page in his mind. For an eidetic imager, mental pictures are remarkably accurate and long lasting.

Almost a century ago, psychologist Francis Galton devised a questionnaire that probed the clarity of people's imaginations. Besides discovering that few people are true eidetic imagers, Galton discovered that people's visualization skills do not improve easily. Here is a modification of his test:

Imagine Last Night's Dinner

Spend a minute or two picturing the scene. Visualize the people you were with, the surroundings, the table settings, the taste of the food, the sounds you heard. Before reading on, take some time to allow your image to build, piece by piece. Start now.

Answer the following questions about your image.

- ❑ Is your image clear or indistinct?
- ❑ Is the image brighter or dimmer than the original scene?
- ❑ Are all parts of the scene sharply defined at the same instant, or are some parts clearer than others?
- ❑ Does the image appear in color or in shades of gray?
- ❑ If your images are in color, are the hues accurate?
- ❑ Can you form a single visual image of the entire dining room?
- ❑ Can you retain a steady image of your dinner plate? If so, does it grow brighter?
- ❑ Can you mentally see your dinner plate, your hands holding a knife and fork, and a person's face sitting across from you all at the same time?
- ❑ Can you feel the texture of the food?
- ❑ Can you picture what people were wearing?

How did you do with this exercise? If you are like most people, you probably found that there are places in your imagery that are rich and vibrant, and places that are less clear. Perhaps you could imagine the people's faces, but not the shape of the cups. Maybe you could recreate the smell of the food but not the taste. Or recreate the acoustics of the room but not recall the shape of the cutlery. Perhaps all you could generate was the *idea* of yesterday's dinner without actually seeing any image at all.

Regardless of how vividly mental pictures can appear, our imagery tends to be far less accurate than we imagine it is. Draw a dollar bill, or your front door, or a telephone dial's digits and letters, or your car's dashboard — things you see every day — and you'll find gaps and holes in your mental images. The accuracy of a mental image depends greatly on how much you have analyzed the original object. If you've never really examined a dollar bill closely, you probably won't be able to form an accurate image of it. If you rarely give your dashboard a second glance, you won't have more than an abstract sense of what it looks like.

To add life to your imagery, you need to sidestep verbal thinking and use your graphic analysis muscles. Say you wanted to create a mental image of the swirl below. What would you do? Begin by asking yourself, what do I actually see? Spend a moment looking at its overall shape. How light or dark is it? What, if anything, does the shape make you feel? What associations do you have with the image? Does it remind you of anything? A car tire? The digit zero? A Cheerio? The Zen symbol for eternity? Now look carefully at the details, focusing on the individual shapes in the figure. Notice how the edges and lines form miniature figures of differing sizes. What are the spatial relationships of these shapes? How do they build together to form the larger shape? Look at the space surrounding the shape, and how the space, itself, defines the shape. Continue to examine the shape in different ways until you get a solid feel for the object. The more you see and know about something, the easier it becomes to form an image.

When you want to form a clear image, recall the key elements gleaned from your analysis — the sense of proportion and the arrangement of component shapes. Allow the image to appear in front of you. It may be difficult at first, if you do not have the habit of visualizing, but if you give yourself time the images will grow brighter and steadier.

What objects can you practice visualizing? Faces, clothes, cars, buildings, company logos, patterns on concrete, pens, pencils, wallpaper designs, book covers, or anything that catches your interest is fair game. Besides encouraging you to pay more attention to your environment, visualizing everyday objects gives you a deeper appreciation of what's happening around you and deepens your sense of reality.

When you read a novel, take an extra moment to picture the events. Visualize the setting, the characters, and the action. Choose a specific spatial viewpoint and let the story unfold in front of you. Similarly, next time you read a newspaper article, visualize what's happening in the article. If you read about a political leader making a statement, picture the politician speaking the words. If you read about an earthquake, imagine what it would feel like to be there. You'll be astonished at how much more you will remember of what you have read.

As you practice visualization, you may find that at times images appear slowly; as in a jigsaw puzzle, they build up piece by piece. At other times, images emerge all at once, spontaneously, in full living color, giving you a taste of how brightly and clearly mental pictures may appear to an eidetic imager. More than anything else though, the practice of forming sharp, accurate mental images encourages you to really *see*.

TIP Take the time to visualize.

A Glance to Your Eyes Is Sufficient

The following assortment of exercises is designed to improve the vividness and the controllability of your images.

❑ **The Real Thing**
Visualize each of the following items. If the images don't appear as bright as you want, don't try to force them into being. Instead, focus on the *idea of seeing an image.* Know that whatever you are trying to visualize has a shape, a texture, a color, and a size. Focus first on the shape and then fill in the details. Take time to allow the image to become steady and sharp.

a familiar face	a childhood friend
a running dog	your bedroom
a sunset	a flying eagle
a babbling brook	a drop of dew
cirrocumulus clouds	a massive oak tree
a typewriter keyboard	a snow-capped peak
a toothbrush	your favorite pair of shoes

❑ **The Not-so-real Thing**
What is the difference between images of things you have seen and images of things you haven't seen? Picture in your mind the following imaginary items.

a unicorn	a chocolate river
a demigod with six arms	a hobbit
a talking giraffe	a thirty-foot ant
a choir of angels	a four-dimensional sphere
World War III	a five-leaf clover

❑ **Quarter and Drawing**
Keeping your eyes still, look to the rim of your vision. Use your peripheral sight to see into the full hemisphere of your vision. Close your eyes and try to reconstruct the image. Mentally divide the view into four quarters. Choose a quadrant and analyze everything in that quadrant, then draw that quadrant. Repeat for each quadrant.

Five Mental Pictures

Visualize five things that are blue. *(blueberries, sky, book cover, ...)*
Do the same for other colors such as red, yellow, green, and purple.
Visualize five things that begin with the letter A. *(artichoke, aardvark, ...)*
Do the same for each letter of the alphabet.
Visualize five things smaller than your finger. *(pen tip, pea, blood cell, ...)*
Visualize five things that are larger than a bus. *(blue whale, train, ...)*
Visualize five things that are found under the ground. *(roots, worms, ...)*
Visualize five things that make you happy *(surfing, ice cream, ...)*

Afterimage

If you close your eyes after looking at an object, you will see an afterimage that lingers for a few moments. Try to incorporate this afterimage into your visualization. For example, look at a pen, close your eyes, and see the afterimage. When the image fades, open your eyes, look at the pen, close your eyes, and see the afterimage. Repeat this a few times at a comfortable pace until you can clearly picture the afterimage, even if for only a second. Then try consciously to create the afterimage of the pen at will.

Peripheral Mental Vision

Visualize a pen in front of you. Then picture it slowly making a circle around you, so that it is to your left, behind you, to your right, and then back in front of you. As the pen travels, imagine that you are using your peripheral vision to see it.

Picturing People

Visualize all the people you have spoken with today. What did they look like? What color were their hair and eyes? What were their heights and ages? What clothes were they wearing? Can you picture their mannerisms and habits? Visualize the people you saw yesterday; last weekend; on a recent holiday; on your last birthday.

In the Eye of the Beholder

Picture a butterfly, perfect in every detail, with gold-piped wings, long furry antennae, and delicate markings, but imagine it being ugly instead of beautiful.

❑ Mathematical Entities

A great deal of mathematics is performed with imagery. Visualize each of the following three-dimensional shapes. Don't try just to form the image: also see the inner structure and relationships between the flat shapes. Mentally manipulate the figures, viewing them from all sides, including the inside. Try to get a sense of the tridimensionality of the solids.

sphere	cube	prism
tetrahedron	pyramid	dodecahedron
octohedron	icosahedron	stellated octohedra

❑ Feelings, Nothing More Than Feelings

Visualize a positive emotion. Mentally picture the feeling of astonishment, without seeing a specific object or memory. Try visualizing a wish, without wishing for anything in particular.How vividly can you visualize the following emotions: hope, joy, fulfillment, love, anger, apathy.

❑ Mental Presence

Form an image of the *presence* of your mother, without actually *seeing* your mother. Similarly, summon a mental image of being in the presence of a mountain, without visualizing the look of the mountain.

❑ The Giant Body

Mentally picture an enormous human body. Imagine flying around this body, seeing it from various points of view from and from various distances. Imagine that the body is so large that you can enter the nostrils, mouth, and other openings, and explore the inside of the body. Make the body so large that you can fly across landscapes of cells, and even view the components of individual cells.

❑ Color Crazy

Visualize the color blue. You may want to begin by visualizing a specific object, like a blue car. Then make the car so large that it fills your entire mental field of vision. Bathe in the color. How bright can you make it? Try picturing other colors: purple, yellow, red, orange, and green. Try imagining color blends; red to blue with intermediate shades in between. Picture red above you, blue to the left, and green to the right.

❏ TV Repairman

Choose a specific object, like a blueberry cheescake, and try to form a clear mental image of it. Imagine that you can adjust your mental image the way you can adjust a television picture by turning various dials.

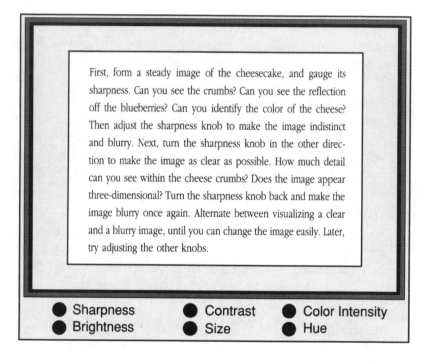

First, form a steady image of the cheesecake, and gauge its sharpness. Can you see the crumbs? Can you see the reflection off the blueberries? Can you identify the color of the cheese? Then adjust the sharpness knob to make the image indistinct and blurry. Next, turn the sharpness knob in the other direction to make the image as clear as possible. How much detail can you see within the cheese crumbs? Does the image appear three-dimensional? Turn the sharpness knob back and make the image blurry once again. Alternate between visualizing a clear and a blurry image, until you can change the image easily. Later, try adjusting the other knobs.

● Sharpness ● Contrast ● Color Intensity
● Brightness ● Size ● Hue

❏ Images of Ideas

Picture the idea of Beauty. Do you see a specific image of something beautiful, or can you have an abstract image of Beauty without seeing a particular visual image? Do you associate sounds, smells, tastes, and sensations with your idea of Beauty? How large or small is the idea? Here are some more abstract concepts to contemplate:

change	order	energy	peace
harmony	communication	reality	illusion

Auditory Imagery

❏ **Mental Music**
Play in your mind's ear the following pieces of music. Hear the melody, the orchestration, and words, if any.
Classical Music: Toccata and Fugue in D minor
Rock Music: Stairway to Heaven, Jumping Jack Flash
Pop Music: Yellow Brick Road, Graceland
Sound Tracks: Star Wars, Superman

❏ **Inner Sounds**
Form an image of each of the following sounds:
five sounds heard in nature (*rain, wind, waterfall*)
five sounds heard in a car (*door closing, moving seat, starting ignition*)
five animal sounds (*dog barking, elephant trumpeting*)
five insect sounds (*housefly, mosquito, cricket chirping*)
five mechanical sounds (*air conditioner, turbine*)
five human sounds (*singing, sniffling, hiccupping, coughing*)

❏ **Abstract Sounds**
Can you mentally hear what the following might sound like:
Bright sounds, clear sounds, muffled sounds, horrible sounds, joyful sounds, golden sounds, smooth sounds, bittersweet sounds, harsh sounds, hollow sounds, sharp sounds, wide sounds.

❏ **Mental Recording**
Snap your fingers and listen very closely to the sound. After a moment, recreate the sound in your mind. Do this several times so you can hear the snap at will. Then change the quality of that sound by mentally altering the location of the sound. Make the sound come from above you, from the left, behind you. Imagine the sound louder, then softer. Imagine it in a small room, like a closet, and in a large room, like a gymnasium. Do the same for other sounds, like voices.

Tactile Imagery

☐ **Inner Sensations**
Relax in your favorite armchair, and visualize yourself washing your hands. Go through every motion, every nuance of movement, every mannerism. Feel the suds, hear the running water, smell the soap, see the little bubbles. Now wash your hands and compare the experience to your visualization.

☐ **More Inner Sensations**
In the same detail as above, visualize taking a shower, getting dressed, making a bed, walking into work in the morning, having a conversation with your best friend, riding the bus or subway, doing your weekly shopping, making dinner, washing dishes, getting into bed.

☐ **Getting a Sense**
Visualize the sensations associated with the following activities:
five sports *(tennis, baseball.)*
five musical activities *(putting on a record, playing guitar)*
five activities at home *(making breakfast, making a bed)*
five business activities *(typing a memo, sitting at a board meeting)*
five kitchen activities *(washing a cake pan, peeling potatoes)*

☐ **Getting a Feel**
Visualize the texture of the following objects:
five different kinds of cloth *(polyester, silk, cotton)*
five building materials *(concrete, wood, tile)*
five plants *(oak tree, rose petals, ivy)*
five objects in a car *(steering wheel, upholstery, tape deck)*

Taste

❑ **Form a clear image of taste of the following foods:**
five meats (*pork, veal, chicken*)
five vegetables (*potatoes, Brussels sprouts*)
five fruits (*banana, orange, kiwi*)
five cheeses (*emmenthal, cheddar, gruyère*)
five of your favorite meals (*maki sushi, borscht with potato perogies*)

Try visualizing the tastes without seeing a mental picture of the food.

❑ **Multi-tastes**
Form the sensation of each of the basic tastes that your palate is sensitive to: sweet, sour, bitter, salty. Combine these qualities together. What would sweet and sour taste like together? Bitter and salty? Sweet, sour, and bitter?

❑ **Imagine the tastes of the following foods mixed together:**
pickles and ice cream   spaghetti and boiled potatoes
cream cheese on white bread   lettuce and Captain Crunch cereal
rice pudding with apples   milk mixed with orange juice

Smell

❑ **Form an image of the smells of the following things:**
five meats (*pork, veal, chicken*)
five vegetables (*potatoes, Brussels sprouts*)
five fruits (*banana, orange, kiwi*)
five cheeses (*emmenthal, cheddar, gruyère*)
five of your favorite meals (*maki sushi, borscht with potato perogies*)
five flowers (*rose, gardenia, dandelion*)
five smells associated with engines (*gasoline, oil*)
five smells associated with various environments (*ocean, pine forest*)
five smells associated with different weather (*humidity, rain*)

Try visualizing the smells without seeing a mental picture of the object.

Putting Imagery to Work

"Imagination is more important than knowledge."

ALBERT EINSTEIN, *physicist*

Olympic decathelete Bruce Jenner kept a hurdle in his living room for four years. Each time he sat on the couch, he would look at the hurdle and mentally jump over it. Golfer Jack Nicklaus runs a mental movie before each shot. In his imagination, he sees not only where he wants the ball to finish, but how the ball gets there.

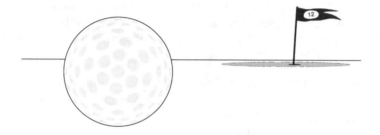

In many ways, the ability to *direct* your imagery can enhance your personal performance. The mental pictures we show ourselves affect our intellects, our feelings, and our bodies — our whole beings.

One reason visualization is so potent is that there is a very strong connection between our mental images and our feelings. Our emotions respond to pictures — of our mothers, a warm cuddly puppy, an attractive member of the oppposite sex — as if the object in the image were actually in front of us. As a result, the mental pictures we see in our mind all day long — images of work and home, of friends and family, of joys and disappointments, of memories and expectations — are closely linked to our outlook on life. In fact, mental images are at the heart of our personal experience: they help create our inner world.

To demonstrate for yourself the power of mental images, try the following exercise:

Emotional Flexibility

Sit down for a few moments, relax,
and visualize a chair
with which you have a neutral association.

When you can form a steady image,
picture the chair in a funny situation.

Imagine it in an outrageous scene
from a Marx Brothers
or Three Stooges movie.

Allow your mind to come up with something bizarre.
You're alone, so there's no need to censor your images.

Do this until you feel refreshed, uplifted, and cheerful.

Now place the same object in a tragic situation.

Center it in a heart-rending scenario.
Be willing to upset yourself.

After you're thoroughly depressed,
go back to the first scenario
and picture the humorous situation
until you've cheered up.

Alternate between scenarios.

Notice the bond
between your mental images
and your feelings.

The Mental Movie

"We are what we imagine ourselves to be."

KURT VONNEGUT, JR., *novelist*

Virtually every moment, most of us experience a steady stream of pictures, sounds, and sensations passing through our mind. This stream of images creates a series of mental movies — thrillers, comedies, horror, love stories — that are complete with commercials, trailers, and advertisments. Because there is a strong link between mental pictures and feelings, negative images affect our moods and interfere with our attention. With deliberate visualization, however, it's possible to alter these images and thereby swing our feelings.

Here are some techniques to apply your visualization skills to life.

- **GOAL VISUALIZATION:** Form a mental picture of yourself doing or having something that you want to have or want to do. You could picture yourself walking in a new pair of shoes, or working at a new job, or eating dinner at your favorite restaurant. In this visualization, three things are important: first, include details by engaging all of your senses; second, imagine yourself *enjoying* the scenario you wish for youself; and third, picture the scene happening on a specific date. By encouraging you to focus on your objectives, this visualization helps you stir up your motivation. Start with small, easy-to-reach goals, and work up to larger, more encompassing goals.

- **PERSONAL PERFORMANCE:** If you are going to give a presentation — in a sales meeting, in a speech, or in an interview — picture a range of possible scenarios. Begin with an ideal outcome. Imagine you're performing in top shape with things happening exactly as you want them to. Then picture a mediocre outcome, with things happening close to the ideal, but not quite. Then visualize the worst possible outcome, with you stumbling over yourself. In the process of rehearsing these scenarios, you prepare yourself for mishaps and problems. Then, once you have worked through the possibilities, focus on the ideal images. Allow yourself to perform marvelously.

82

- **RELAXATION AND STRESS MANAGEMENT:** Take some time to recall pleasant memories. Relax your body as fully as possible, and let your memories float freely through your mind. Touch lightly on great heartfelt memories: green valleys, close friends, early childhood experiences, your old room, your first friend, walking to school in the sun, snow or rain, waking up refreshed and looking forward to the rest of the day.

- **INNER MASSAGE:** Visualize a blackboard. In the upper left hand corner, watch the letters of your name appear slowly, one at a time. Then, in the upper right hand corner of the blackboard, watch the letters of the word "relax" appear one at a time. Pause for a moment, and look at the words. Then, one line down, watch the word appear again, just as slowly.

- **CALMING THE EMOTIONS:** Swami Dyananda Sariswata, a Sanskrit teacher, has an interesting application for mental imagery. To help let go of the inner agitation associated with a troubling situation, focus on the positive. For example, if you are angry at your boss, visualize your boss doing something you like. Recall a time when she did something you approved of. Keep looking for the positive. Then, when you start feeling good about your boss, and thinking that she is not that bad, picture the situation that made you angry, *but in the light of the positive feeling.* Be willing to *accept* your boss as she is. Though you may not *approve* of all of her actions, don't feel compelled to change her or to mentally mar her. Apply this technique to domestic quarrels.

Imagery Style Tips

Mental imagery injects freshness, vitality, and life into what would otherwise be an austere world of abstraction. Make it a habit to explore the vast landscapes and spectacles that exist within your mind.

TIP ONE Train your imagination to be more responsive. Strive to make steady carbon copies of what you hear, see, smell, taste, and sense.

TIP TWO Encourage your inner senses to become more vivid. Step beyond humdrum imagery and encourage your mind to see the colorful, the bizarre, and the unexpected. Be willing to make your inner world as large as you want it to be.

TIP THREE Pay attention to the everday mental images that flow through your mind. Notice what imagery gets you down and what keeps you inspired. Be aware of your power to rescript your mental movie when you want to change it.

"He who has imagination without learning
has wings and no feet."
JOSEPH JOUBERT, *18th-century essayist*

5

MENTAL GYMNASTICS (II)

Thinking with Words

Word Exercise

Imagine that you are a character in a novel.
And imagine that
what is happening around you
right now
is a scene
from one of the chapters.

Spend a few minutes
using words
to describe
what's happening.

Mental Music

*"By words the mind is excited
and the spirit elated."*

ARISTOPHANES, *5th-century B.C. dramatist*

If your mind is a musical instrument, then words — those you think silently and speak aloud — are musical notes that harmonize into melodies and symphonies of thought.

In our heads, in our speech, and in our writing, words give shape and substance to our thoughts. Without words our thinking would lose its richness and intricacy. We would have no way to formulate and express meanings.

Improving your ability to command words involves developing an inner voice. What characterizes a well-trained voice is not so much a powerful vocabulary, but the ability to articulate. The abilities to express words clearly, to infuse them with accents of meaning, and to put feelings into words are signs of a sharpened mind. Keep your inner voice trained and limber by paying attention to how you shape your words and your mind will make beautiful music.

Practice Scales

EXERCISE *Read the following phrase silently to yourself. Sound the words in your mind in the way that a violinist plays a musical phrase on a fiddle. Listen carefully to the quality of the inner music. Repeat the phrase half a dozen times, each time striving to make it a little clearer.*

How clear do the words sound? Is the thought sharp and distinct? Are there subtle breaks in the thought, tiny pauses or hesitations? Are the words rushed because some part of your mind wants to think about something else? Is each word well formed in your consciousness? Do other melodies — thoughts that judge or comment on this first thought — interfere?

Every musician knows that only practice improves musical ability. By repeating songs until they sound just right, practicing scales over and over, jamming with other musicians, and exploring different types of music — jazz, country, classical, pop — musical savvy grows. Why not apply the same techniques to verbal thinking?

Sound the thought at various speeds. Play the thought at your regular rate, whatever that may happen to be. Now, sound the words again, only faster. Do it again, double time. Find out how fast you can play the thought without having the words slur into each other. Can you play it in two seconds? One second? A fifth of a second? Now, slow down the thought. Can you stretch it out to ten seconds? Thirty seconds? A full minute?

Play the phrase in your mind at different volumes. Sound the words louder. Make them as loud as a symphony, or a Boeing 747. Now, turn the volume knob in the other direction and repeat the phrase more quietly. Play the thought so quietly that you can barely hear it above the background noise in your mind.

Sound the words and focus intently on their meaning. Pay attention to the significance of the words. Connect the words together, and try to understand the gist of the idea. Repeat the thought, emphasizing only one word in the phrase at a time. For example, sound the phrase as "I *think,* therefore I am." Do this for each word.

Sound the word in different voices. Form the words in your mind as if someone else were speaking them. Try Peter Ustinov or Woody Allen or Ronald Reagan or your father. Try sounding them as if they were spoken by someone you respect, then someone you don't respect, someone you like and someone you don't like.

Sound the thought in different locations. Place the thought in the middle of your brain. On your forehead. In the back of your head. In your chest. At your feet. To your left. Two feet in front of you. At the far side of the room. In the air all around you.

Picture the words in your mind without hearing them. Visualize the phrase as if it were written on a blackboard. Try focusing on the content of the thought, without hearing or seeing the words.

Sound the thought without trying to act upon it. Imagine the thought is floating as a puff of smoke in still air. Don't manipulate it, just let it float.

Musical Phrases

Here are a collection of quotations to limber up your verbal muscles. Choose one and give yourself plenty of time to ponder its inner meaning. Play the thought through until you can sound it clearly in your mind, with just the right emphasis.

A word to the wise is sufficient. — *Titus Plautus*

To those who are awake, there is one ordered universe common to all, whereas in sleep each man turns away from this world to one of his own. — *Heraclitus*

One great use of words is to hide our thoughts. — *Voltaire*

A man may dwell so long upon a thought that it may take him prisoner. — *Sir George Savile*

The only way to get rid of temptation is to yield to it. — *Oscar Wilde*

Nothing comes of doing nothing. — *William Shakespeare*

The most incomprehensible thing about the world is that it is comprehensible. — *Albert Einstein*

Only the ephemeral is of lasting value. — *Ionesco*

There is nothing so unthinkable as thought unless it be the entire absence of thought. — *Samuel Butler*

Every decision you make is a mistake. — *Edward Dahlberg*

The word "water" is itself undrinkable, and the formula H_2O will not float a ship. — *Alan Watts*

It is completely unimportant. That is why it is so interesting. — *Agatha Christie*

When they tell you to grow up, they mean stop growing. — *Tom Robbins*

It takes a very long time to become young. — *Pablo Picasso*

The present is the only thing that has no end. — *Erwin Schrödinger*

Freedom of will is the ability to do gladly that which I must do. — *Carl Jung*

The Written Word

*"Read not to contradict and confute;
nor to believe and take for granted;
nor to find talk and discourse;
but to weigh and consider.*

*Some books are to be tasted,
others to be swallowed,
and some few to be chewed and digested;
that is, some books are to be read only in parts;
others to be read but not curiously;
and some few to be read wholly,
and with diligence and attention."*

FRANCIS BACON, *17th-century philosopher*

In a sense, there are two different kinds of reading: reading for entertainment and reading for knowledge. The purpose of reading for entertainment is to immerse yourself in another world, sometimes one of fantasy, to escape from day-to-day living. You read to become emotionally involved, interested, and inspired. The purpose of reading for knowledge is to learn more about the world of matter and ideas. You read to collect information, to evaluate, assess, and think actively about a subject.

Because each kind of reading exercises a specific set of mental muscles, it's important to know why you are reading. Do you want to relax and keep yourself entertained? Do you want to skim and just become acquainted with the material? Do you want to add to your storehouse of facts? Do you want to remember everything you are about to read? Do you want to think more deeply about and ponder an issue? When you know *why* you're reading, you'll know better *how* to read.

While reading for entertainment, encourage your imagination to become as vivid as possible. Take time to picture the scenes, to visualize the characters, to feel the emotions. If you read mysteries, exercise your imagination by making up your own ending, or even three or four possible endings. Playwright George Bernard Shaw used to write an outline of each book he read *before* he cracked it open.

Reading for in-depth knowledge usually demands a more deliberate focusing of attention. When you read to achieve a solid understanding, give yourself definite objectives for both the amount and the quality of your reading. Put yourself in a quiet place where it's easy to concentrate. Remove any external distractions — turn off the radio or television — and then settle into a comfortable — but not too comfortable — alert posture.

Very quickly skim over the material and determine how far you want to read. Pay attention to headings and topic sentences. Be on the lookout for key points that establish the logical structure of the presentation. To help remember the material, assign a key word or phrase to each paragraph. Jot down key words either to the side of the paragraphs or on a separate piece of paper. These words should summarize the main idea of the paragraph. Later, if you read through a list of key words, you quickly refresh your memory.

Besides picking out key ideas, take some time to evaluate and weigh ideas. Are the main points supported? What evidence leads to the conclusions? What might be missing from the presentation? Rethink the subject in your own terms, putting it into your own logical framework. It's always easier to remember your own words than someone else's.

Build good reading style. Avoid the tendency to skip over words that you don't fully understand. Unknown words tend to clog the flow of our comprehension, so be willing to take the time to consult a dictionary. You may be surprised to discover that you don't know the accurate meanings of words with which you are familiar.

Avoid subvocal speech. When reading for speed and comprehension, don't sound words in your head or move your lips as you read. Most reading experts claim that for top performance, you need to click your brain into higher gear by focusing on ideas, images, and meanings, rather than on the sounds of words.

Pay attention to how your eyes move across a page. Slow readers tend to make many small jumps, picking out only one or two words with each fixation. Fast readers tend to make fewer and more widely spaced jumps. They take in more with each glance. As a result, they tend to read *down* a page, rather than *across* a page. As an experiment, next time you read a newspaper, try to take in more with each glance of your eye. Don't rush, simply allow your eyes to see with a wider angle.

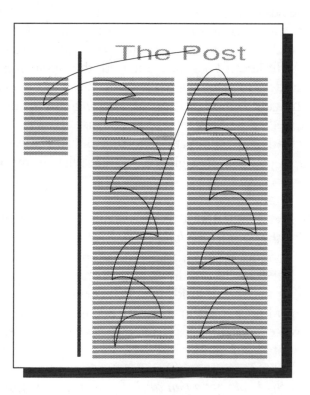

Letting Words Flow

"I write to know what I think."

GRAFFITO

If reading exercises your mind by making you analyze and synthesize ideas, writing exercises your mind by making you clearly formulate and express ideas. An innovative approach to help you think with words, developed by author Gabriele Lusser Rico, is to organize words in a visual tree diagram. Rico called the technique clustering.

Clustering allows your mind to wander freely around concepts. Because the structure is open-ended, you can add ideas anywhere. Like a plant growing outwards, the tree structure encourages ideas to take root, develop, and branch out. While the presence of one central thought keeps your ideas focused, the branching structure allows your ideas to reach out freely. You splash ideas onto a page, writing as quickly as thoughts appear, keeping closely related ideas clustered together, and distantly related ideas separate. The various ideas are clearly separate, yet connected. Because the lines show relationships, you can, in a glance, see the relative importance of each.

EXERCISE Get a pen and a piece of paper. In the center of the paper, write the word *Freedom*. Draw a circle around it, and jot down any thoughts, feelings, and associations that you have about the idea of freedom. Record these thoughts by drawing a circle and writing one or two words inside the circle. Connect related ideas by drawing lines between them. Let your ideas branch out in every direction. Write down the ideas as quickly as you can until no more ideas come into your mind.

Put down the book, get a pen and paper, and start now.

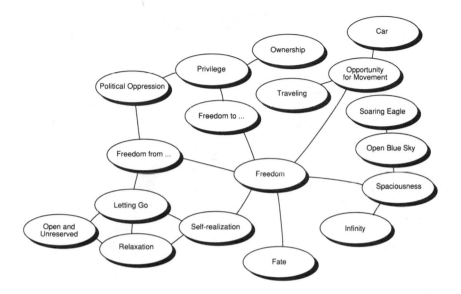

Clustering can be an end in itself — a way to think about what you want to think about — and it can be a means to organize your thoughts — to write, to plan out projects, to explore the options in a decision, to take notes, or to study a subject. Clustering encourages you to switch direction of thought quickly and effortlessly.

Next time you need to write out your thoughts, use the clustering technique. For practice, try clustering the following subjects: the arms race, current fashion trends, the uses of wood, how to solve domestic quarrels, the reason history repeats itself, television commercials.

"A word is a storm center
of meanings, sounds, and associations,
radiating out indefinitely
like ripples in a pool."
NORTHROP FRYE, *biblical scholar*

Word Exercises

❑ **Circle of Thought**

How many words can you find in this circle of letters? Start at any letter, and travel in a clockwise direction.

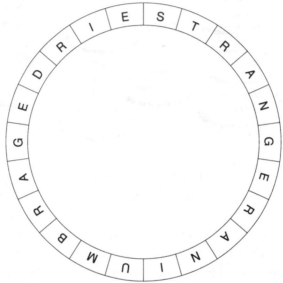

❑ **Conversations**

An old adage says, "The best conversations are those in which you respond to ideas, not words." Tomorrow in a conversation, mentally repeat what the other person says before you respond to him or her. Get a sense of what the person is really saying.

❑ **Word Templates**

See how many meaningful phrases you can make up in which the first letters of each word put together spell *idea*. For example, Individual Demands Endanger Arbitration. Choose a subject, such as the history of the church, and create your sentences around it. Here are some other template words.

YES	BIG	STINK
SOLO	TIP	MOVE
THINK	FEEL	INSPIRE

96

❏ **Spontaneous Speechmaking**

As soon as you read or hear each of the following words, make four statements aloud about the word. This exercise is best done with other people.

silly	bats	sugar	liquid
binder	key	camera	grid
record	program	money	sex
sink	tiger	travel	reality

❏ **Can You Say It Another Way?**

Human beings, who have the gift of speaking, have at their disposal a rich and varied vocabulary, as well as the knowledge of where and when to use words. They can find the word that expresses the exact shade of meaning they wish to convey. How many words or phrases can you come up with that mean the same or almost the same as the following words?

Foolish: *dumb, idiot, dimwitted, short sighted ...*
Important: *weighty, eventful, major, significant ...*
Sad: *somber, down, blue, melancholic ...*
Friend: *comrade, chum, mate, companion ...*
Fearful: *timid, anxious, lily-livered, yellow ...*
Humorous: *funny, a riot, witty ...*
Attractive: *alluring, fascinating, appealing ...*
Self-assured: *cool, confident, brassy, certain ...*

❏ **The Full Spectrum**

Test your ability to come up with words by compiling a list of twenty-six related words in alphabetical order.

Occupations: *architect, baker, chemist ...*
Cities: *Albuquerque, Brussels, Cairo ...*
Musical Instruments: *accordion, banjo, cello ...*
Animals: *aardvark, buzzard, crane ...*
Foods: *artichoke, bread, cranberries ...*

Compile similar lists from your own special field of interest, such as names of plants, insects, pop songs, or streets.

Naming Things
Look around the room you're now in and choose five objects. Give them alternative nonsense names. For example, a pair of eye glasses might be called "bitshifters" or "prangees" or "calmerts." Try to come up with sounds that suit the shape of the object.

Is That What That Means?
English contains many unusual words that are rich in meaning.

comate — hairy or shaggy
pysosis — the formation of pus
seely — arousing contemptuous pity because of weakness
talion — retaliation in the sense of "an eye for an eye"
glossa — the tongue of an insect
clepe — an archaic term meaning to summon by name
layboy — a machine that stacks paper into even piles

Invent meanings for the following nonsense words:
crustophenpalious   rampithicrate
illuspronatural   carceptinous
pharalpleupeian   harmplot
tinbucker   presstumb
bleesh   emoushe
shebarrah   kagnst

Descriptions
How would you describe the following things to someone who is entirely unfamiliar with them:
snow
peanut butter
how to put on a jacket
sexual intercourse
the satisfaction of raking a lawn free from leaves
what your place of work looks like
a Bach fugue
mud pies
baseball

❏ Descriptions
Describe the following shapes to someone who cannot see them. Have that person draw a picture of the shapes from your description.

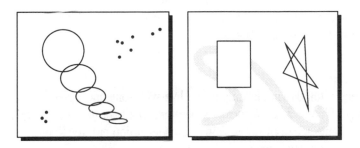

❏ Make up a limerick using one of the following first lines:
There once was a Vicar named Gust ...
A traveling salesman rang twice ...
A girl who weighed much too much ...
There was a smart lad from Madras ...

❏ The Feel of Words
There seems to be the right name for just about everything. Test your ability to create innovative names that grab people's attention. Make up five names for:

a perfume
an introduction service
a pop song
a pet dog

a fashion line
a wheelbarrow
a jazz band
a computer company

❏ Funny Words
Make up a joke from the following subjects:

a dog in the White House
a traveling salesman in a mortuary
a political leader's name
knapsacks
a fisherman in front of the gates of St. Peter

❏ **Intentional Illiteracy**

Take an hour out of your day and consciously become illiterate. If you're walking along a city street, look at the shop signs, look at the labels and advertisements. Don't form the meaning of the words. Instead, look at the shapes of the letters as you would examine the letters in a foreign alphabet. If nothing else, you will certainly become aware of the incredible volume of words that surrounds you.

❏ **Fasting**

Spend a full day not talking. If you live with someone, agree beforehand that neither of you will speak to the other. Get your messages across with body language or by notes. Don't watch television or listen to music or the radio. For a day, keep yourself free from words.

❏ **Story Creation**

At random, choose one word from each of the four columns, and use them as trigger concepts to construct a story. Either in your mind, or on paper, develop the scenario.

fizzle	ocean	travel	sandwich
wallet	watermelon	dream	motorcycle
sideline	criminal	statue	toenail
pavement	topcoat	veneer	charisma
army	nose	Frisbee	spiral
finish	button	soufflé	box
stick	steal	fruit	cowboy

❏ **Tongue Twisters**

Recite the following tongue twisters as quickly as you can:

What noise annoys a noisy oyster? A noisy noise annoys a noisy oyster.
Red leather, yellow leather.
A proper cup of coffee from a proper copper coffee pot.
The sixth sheik's sixth sheep's sick.
Lemon liniment.
I can think of six thin things and of six thick things too!

The Mental Monologue

"Each of us is a singular narrative,
which is constructed continually, unconsciously, ...
through our perceptions, our feelings,
our thoughts, our actions; ...
A man needs such a narrative, ...
to maintain his identity, his self.
If we wish to know about a man,
we ask what is his story."

OLIVER SACKS, *neurologist*

If you're like most people, you probably hear an almost never-ending stream of words flowing through your mind. You rehearse conversations, you formulate, state and re-state opinions, you rate yourself and other people, and you make comments on what's happening. Like a radio playing in the background, these words color your inner environment.

Together with the mental movie, your mental monologue shapes your outlook on life. If you tell yourself that you're a nincompoop, you'll start behaving like one. If you talk to yourself about things you would have rather done, you'll live in the past.

Sometimes the mental monologue can get out of hand. When you're agitated, the words flow through your mind and it's as if you don't have any control. Other times, you think you are in control until you try to quiet the words. Then you realize you are hearing your old mental habits, a relic of what you've said to yourself before.

To explore how you think with words throughout the day, consider some of the following ideas.

- **THE SENATE:** Imagine that the words you hear in your mind don't come from a single speaker but from a senate of speakers. Members of the senate can represent the opinions of your parents, teachers, friends, as well as opinions expressed in books or magazines, and on television. Next time a loud senator dominates the floor, put the issue up for debate. For example, say that a senator announces that there are too many immigrants entering the country. Take a few minutes to let the other, perhaps less vocal, speakers state their opinion. Try to identify who the senators represent and let all the senators have their chance to speak. Let yourself be willing to entertain several points of view at once.

- **TAKE TIME TO LISTEN:** It's useful to give yourself time to slow down and really listen to what you're saying to yourself. One way to do this is to visualize a portable radio and the words that are coming from the radio. It's a very high quality radio so you can hear the voices behind the voices. When you give the mental monologue a location, it's easier not to get caught up and distracted by it. Remember, to open the door to deeper stages of thinking, the key is to let go and be sensitive.

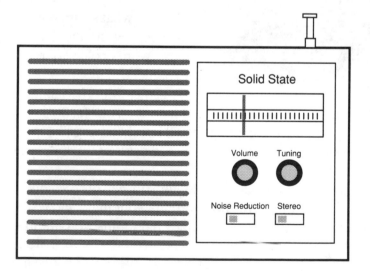

- **RELABELING:** In our mental monologue, we tend to label our experience — we categorize it. Say that you need to speak in front of a large audience and your palms become sweaty, your stomach begins to make strange sounds, and your face turns a little flushed. You could tell yourself that you feel nervous, but that would probably make you more nervous. Or you could tell yourself that you were feeling the patterns of energy that naturally accompany important situations. That would enable you to redirect that energy into your talk. If you feel butterflies in your stomach, fly them in formation.

- **GETTING RID OF MIND BABBLE:** Perhaps you hear a voice in your mind repeating negative thoughts. You might hear that you're a dummy, and that you will never amount to anything. Repeat the phrase in a high, squeaky voice. Use the same words, just change their quality. Or play the song "Turkey in the Straw" over the voice. If you're in a rush, and a speaker says, "You're late. You're late," slow down the voice to a crawl and see what happens. If you hear a particularly annoying voice in your mind, change the quality of the voice. Get control over it by making it louder, then softer, faster, then slower, closer, then more distant.

- **POWER MANTRAS:** A mantra is a word or phrase that when repeated brings about a specific mental state. To make a mantra work for you, spend a few moments relaxing and then sound the word in your mind, which will trigger the associations you want. For example, sound the word *relax* in your mind and you will relax. The trick is to find a rhythm that lets you concentrate. Make each word a complete thought separated by a moment of silence. Listen to the silence in your mind after the word.

Word Style Tips

Words are the exposed surface of thoughts. Develop your inner voice to become more versatile in the way you express your thoughts to yourself and to others.

TIP ONE Pay attention to the quality of your thoughts. Improve your mental articulation by varying the tempo of your thoughts. Slow your thoughts down. Speed them up.

TIP TWO Make time to read. Reading is a powerful mental exercise that encourages your attention to become steadier and your imagination to become more vivid.

TIP THREE Put your thoughts down on paper. Writing encourages you to sort out and develop various points of view, to establish and challenge opinions, to explore what you feel, and to think about what you want to think about.

TIP FOUR Pay attention to what you tell yourself. Listen to your mental monologue and discover how the messages that pass through your mind all day long affect your outlook. What is your inner story?

"A word is not a crystal, transparent and unchanging,
it is the skin of a living thought
and may vary greatly in color and content
according to the circumstances in which it is used."
OLIVER WENDELL HOLMES, *19th-century jurist*

6

MENTAL STRENGTH

Problem Solving

Time Is of the Essence

You are shipwrecked and captured by cannibals.

Two hourglasses are thrust into your hands.
One measures exactly four minutes.
The other measures exactly seven minutes.
The chief cannibal says you must tell
when exactly nine minutes have passed.

If you can do this,
you may go free.

If you can't,
you're dinner.

The chief yells, "Start now!"

What do you do?

Inner Challenges

*"To live is to have problems,
and to solve problems
is to grow intellectually."*

J.P. GUILDFORD, *psychologist*

How did you do with the cannibal-hourglass problem? When this puzzle was first told me by a good friend, he thrust two imaginary hourglasses into my hands. I sat down and began to think about what to do. After a couple of moments, my friend told me, "Too late. You missed your chance. You're now part of a stew." I looked up and said, "What do you mean? I haven't told you what I'd do yet." He said, "That's just the point. When you tried to think about a solution, your opportunity slipped away. You held the hourglasses in your hand without turning them over. Since you took no action, you would have started too late, even if you figured out how to time out nine minutes. When the chief said, 'Start now!', he meant start now!"

At first I felt cheated. But then I realized that there was a lesson to be learned. From my friend's dramatic presentation, I realized that I hadn't responded to the problem as if it were really happening. I hadn't put myself within the situation. I had tried to solve it from outside.

How would you solve the problem, if you really were shipwrecked? The first thing is to turn over both hourglasses. Now, as the sand is pouring through, you have time to think. What happens next? After four minutes, the first hourglass runs out, and you do the natural thing: turn it over. At seven minutes, the second glass runs out. Turn it over. One minute later, at eight minutes, the first glass runs out once more. Since you have one minute more to go, and since you also have exactly one minute of sand in the bottom of the second glass, you can turn over the second glass when it's only partially filled. When the one minute of sand runs out, you're at nine minutes. Just in time.

Good problem solving involves developing a basic set of skills that can be applied to virtually every kind of problem — abstract problems, relationship problems, money problems, problems at work. Some of these skills include:

- **DIRECT APPROACH:** When you encounter a problem, do you move towards it or shy away from it? Are you satisfied with one solution or do you try to find many alternative solutions? Do you skip over or hurry through problems or do you confront them directly?

- **ORGANIZATION:** To solve problems effectively, you need to organize information in a clear and manageable way, zeroing in on key factors and important relationships, and ignoring extra baggage. You must define the problem exactly. Einstein understood the importance of clarity when he said, "Everything should be made as simple as possible, but not simpler."

- **MANIPULATION:** Some problems are solved by a well-defined sequence of individual steps. Other problems are solved through experiment and trial and error. Still others are solved only through lateral thinking, and the answer suddenly arrives in a flash of insight. In each case, you must manipulate the information. To do this you may rearrange words, or see pictures, or think in mathematical symbols, and finally you check your gut reaction.

- **TESTING:** Once you arrive at a solution, check to see if your solution is correct. Re-examine your assumptions, and confirm your train of reasoning. Ask yourself if there is another way to arrive at the solution.

There is no foolproof formula that guarantees a solution for every kind of problem, since different kinds of problems require different kinds of thinking. The following section contains some classic puzzles. You may discover that many of the solutions have elements of play, humor, and beauty. Look in the **End Stuff** section to find the solutions. Remember to incorporate the four strategies into your own approaches. With time, you will acquire insights you can apply to many other problems.

Inner Arranging

If three days ago was the day before Monday, what will the day after tomorrow be?

The first step towards solving any problem is to figure out what you need to figure out. When you form an idea of what you are looking for, you have a goal for your thinking, and you have a place to start. In the first puzzle, the objective is to find out what the day after tomorrow will be. With this in mind, you can restate the puzzle in simpler terms. The day before Monday is Sunday. If three days ago was Sunday, then today is Wednesday. If today is Wednesday, then the day after tomorrow is Friday. The puzzle becomes almost trivial when you arrange the information clearly.

In the following puzzles, start off by determining your objective and then restate the puzzle in a way that makes it easier to approach.

❑   A zoo in Tallahassee has thirty heads and one hundred feet. How many beasts and how many birds does the zoo contain?

❑   Three missionaries and three cannibals need to cross a river in a boat that can hold only two people. The missionaries must be careful to never allow the cannibals to outnumber them on either shore. How can everyone cross the river?

In the following puzzles, each letter represents a different digit. Can you decode the correct letter-number combinations?

```
  S E V E N          S E N D          A B C D E
-   N I N E        + M O R E          x       4
  E I G H T        M O N E Y          E D C B A


    W I R E          W R O N G        L E T T E R S
  + M O R E        + W R O N G      + A L P H A B E T
  M O N E Y          R I G H T        S C R A B B L E
```

In a game called flip it or dip it, three people flip coins, and the odd person out has to pay the others double what they have already won. After three games, each person has lost twice and has thirty-six dollars. How much did each start with?

Five hundred women live in a remote village in Papua New Guinea. Six percent of them wear one earring. Of the other ninety-four percent, half are wearing two earrings, half are wearing none. How many earrings are there in total?

Stan and Ollie enter a restaurant. Stan orders a black coffee, and Ollie orders a cup of milk. To satisfy their tastes, Stan takes a tablespoon of Ollie's milk and stirs it into his coffee. Then Ollie takes a tablespoon of Stan's mixture and stirs it into his milk. Is there more milk in the coffee than coffee in the milk, or are the proportions the same?

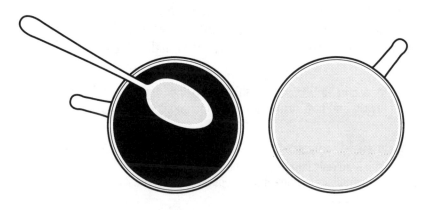

Picturing

One good way to solve problems is to draw pictures of them. A visual representation, such as a simple sketch or diagram, can force the problem to hold still so that you have time to work around it. The skilled logical thinker realizes the importance of conceptualizing the problem graphically. With pen and paper, you can write down all that you know about a problem and organize the information into a cohesive whole. Here is an example:

One morning, exactly at sunrise, a monk began to climb a tall mountain. The monk followed a narrow path and stopped many times along the route. Exactly at sunset, he reached the summit. After a night of meditation, he started his journey back exactly at sunrise. As before, the monk paused along the path many times, reaching his destination, the mountain base, exactly at sunset. Prove that there is a single spot along the path which the monk will occupy at the same time of day on both trips.

You can solve this problem by drawing a picture of the situation. Imagine that the monk's path is spread out on a time–height graph. When you superimpose both of the monk's journeys, the answer becomes clear: the monk occupied a single position on the mountain at the place where the two lines representing his path intersect.

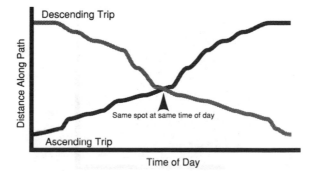

❏ A modern artist has ten identical statues that are to be exhibited in one room of your gallery. The artist insists that against each of the four walls, there must three statues. How would you arrange the figures?

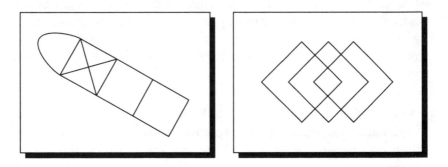

❏ Try to draw the following figures in one line without lifting your pen or pencil from the paper.

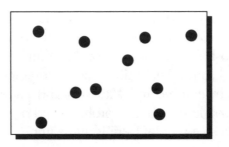

❏ Can you fold a square piece of paper to form the creases of a regular hegaxon? You may not use a pencil or ruler. The hexagon can be in any position in the square.

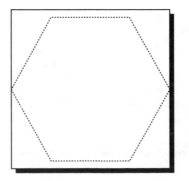

Finding Patterns

There is a well known story about mathematician Karl Freidrich Gauss. As a schoolboy, Gauss was given an assignment to find the sum of the numbers from 1 to 100. The teacher, hoping to keep the class busy for as long as possible, was more than a little surprised when Gauss had the correct answer in only a few moments. Gauss had realized that the numbers could be arranged in pairs which add up to one hundred. For example 1 + 99 = 100, 2 + 98 = 100, 3 + 97 = 100. Since there are forty-nine of these pairs, plus the numbers 100 and 50, the grand total is 5050.

Many problems can be solved by finding a hidden pattern. Discovering such a pattern involves stepping back and looking at the problem from a distance. See if you can find the short cuts in the following puzzles.

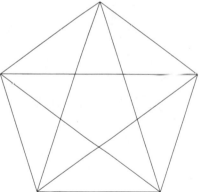

❑   How many triangles can you find in this figure?

❑   Seven men and two boys need to cross a river. The one canoe is tiny and can carry either one man or two boys. How many times does the boat have to cross the river for everyone to get to the other side?

❑   A mouse is trained to negotiate a maze so that it takes a path where each turn takes it closer to the cheese. In the following maze, how many possible paths are there?

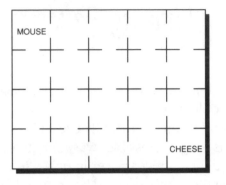

☐ Place the numbers from 1 to 19 in the 19 circles so that any three in a straight row add up to 30 with the center circle always being in the middle of the row.

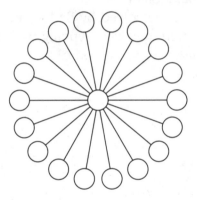

☐ How many ways is it possible to read the word RADAR in this diagram? You may travel in any direction and use every letter along the path.

114

Reasoning

If it takes Big Ben thirty seconds to chime six o'clock, how long does it take to chime twelve o'clock?

Sometimes, the solution to a problem can seem obvious, when it is not. The solution to the above problem is sixty-six seconds. When Big Ben strikes six o'clock, there are five intervals between the first bell and the last bell. The time between each interval is six seconds (one fifth of thirty seconds). When Big Ben strikes twelve o'clock, there are eleven intervals between the first and the last gong. Since the interval is six seconds long, it takes Big Ben six times eleven or sixty-six seconds to chime twelve o'clock.

Try your hand at these reasoning and deduction problems.

❑ Three cookie jars are incorrectly labeled as "Oatmeal," "Peanut Butter," and "Peanut Butter and Oatmeal." The jars are closed, so you can't see inside. You must take one cookie from only one jar, and then correctly label each jar. Which jar do you take the cookie from?

❑ As a birthday gift, you receive five identical gold chains, each containing four links. You want to combine them into a single necklace. If it costs a nickel to open a link, and seven cents to close a link, how little do you need to spend to get a single full chain?

❑ It's midnight and there's a power failure. You have twenty-two green socks and thirty-five purple socks in your dresser drawer. How many socks do you have to take out to be sure of having a pair that matches?

❑ In Kurdistan, money is denominated in lengths of silver. A repairman took fifteen days to fix a person's house and wanted one inch of silver at the end of every day. The house owner, who had a bar of silver fifteen inches long, devised an ingenious method of payment that used only four cuts. How did this work?

❑ Your eccentric Uncle Jake is a rare coin collector. He has twenty-four coins that appear identical, except only one is solid gold; the others are made of a heavier alloy. He sits you in front of a balance and says you can have the gold coin if you can find it. You can use the scale only three times. How can you find the lightest coin?

❑ You come to a fork in the road to Tipperary and meet two men. You know that one man always tells the truth and the other always lies. But you don't know who is the liar and who is the truth teller. To get to the right destination, you can ask only one of the men one question. What should that question be?

What Are Your Assumptions?

Connect the nine points together with four straight lines without lifting your pen or pencil from the paper.

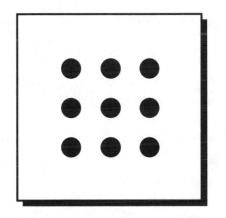

You may have made several assumptions about the problem that limited your range of answers. One assumption that many people make is that the lines may not extend beyond the imaginary square formed by the box. Break that assumption and you can solve the puzzle more easily.

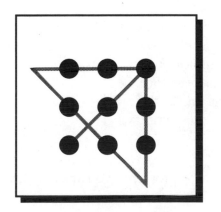

If you break other assumptions, you find more solutions.

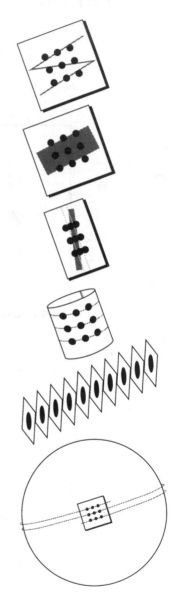

- ASSUMPTION: *The lines must pass through the center of the dots.* If you draw lines that just touch the dots, you can solve the puzzle in three strokes.

- ASSUMPTION: *The lines must be thin.* Connect the dots with one fat line and you have solved the problem.

- ASSUMPTION: *You may not crease the paper.* Fold the paper in such a way that the dots line up. In this case, you only need one stroke.

- ASSUMPTION: *The paper must be flat.* Roll the paper into a tube and it's possible to connect the dots with a spiral. When unrolled, the spiral becomes a straight line.

- ASSUMPTION: *You cannot rip the paper.* Poke a hole through each of the dots with your pencil, and the dots are all connected up.

- ASSUMPTION: *The lines cannot extend beyond the edge of the paper.* If a line is long enough, it travels around the earth and returns from the opposite direction. One line that circumnavigates our planet twice will solve the problem.

To solve the following problems, you must think beyond some usual assumptions.

❏ With six matches, construct four equilateral triangles. You may not bend or break any of the matches.

❏ Move one and only one of the four matches to make a square. You may not bend or break any of the matches.

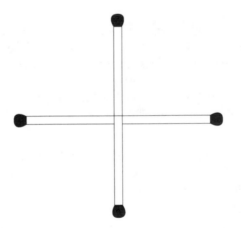

❏ Two ropes of rare silk are hanging in a monastery. A thief, who happens to be an acrobat, wants to steal as much of the ropes as he possibly can. The ropes are three feet apart and fixed in the middle of a sixty-foot-high ceiling. The thief knows that if he jumps or falls more than fifteen feet, he will not be able to walk out of the monastery. Since it's not possible to bring in a ladder, he must climb the ropes. After a great deal of thought, the thief figures out a way to steal almost all of the ropes. Can you figure it out too?

Lateral Thinking

One of my favorite kinds of puzzles is that in which the answers arrive in a sudden flash of insight. Martin Gardner, mathematician and puzzle aficionado extraordinaire, calls this class of puzzles "Aha!" puzzles.

Try your hand at solving some of these leaps of logic:

❏ Jane hailed a cab. Along the way, she talked so much that the taxi driver became irritated. He said to Jane that he was sorry that he couldn't hear a word she was saying. He added that since his hearing aid wasn't working, he was deaf as a doorknob. Jane stopped talking, but after she arrived at her destination, she realized that the cabbie had pulled a fast one on her. How did she figure out the cabbie was lying?

❏ An evil moneylender offered to settle a debt with a young woman if the woman would agree to a simple wager. She was to reach into a bag and draw out one of two pebbles. If the pebble was white, the debt would be forgotten. If it was black, she would have to marry the moneylender. The young woman noticed, however, that the moneylender put two black pebbles into the bag. What did she do?

❏ Frank was sleeping in an anchored ocean liner. At noon the water was six meters below the porthole and was rising one meter per hour. Assuming that this rate doubles every hour, when will the water reach the porthole?

❏ Two sets of train tracks run parallel except when they enter a tunnel. The tunnel is too narrow to accommodate both sets of tracks, so for its entire length, the two become one. One afternoon, one train entered the tunnel from the south end, and another train entered from the north end. Both trains were travelling at top speed, in opposite directions, but there was no crash. Why?

❏ Why are 1988 dollar bills worth more than 1987 dollar bills?

☐ A new type of bacteria has been discovered that divides into two every hour. Each of the pair splits into four the next hour. Eight bacteria are placed into a Petri dish exactly at nine A.M. At midnight the dish is completely filled. At what time is the container exactly one quarter full?

☐ A man kisses his wife before he leaves for work. He shuts the apartment door, walks into the elevator, presses the ground floor button, and realizes instantly that his wife has just died. What happened?

☐ A sailor walks into a restaurant, sits down at a table, and orders albatross. When the meal arrives, he takes one bite, leaves the restaurant, takes out a gun, and commits suicide. Why?

☐ A woman is arrested for murder. She is tried, found guilty, and sentenced to death. However, the execution can never be carried out. Why not?

☐ A woman enters a room and gets a glass of water. After drinking the water, she holds her breath for half a minute. Out of the corner of her eye, in the reflection of a mirror, she sees someone behind her with knife in hand, ready to stab her. She screams, the man lowers the knife, and they laugh together. What's happening?

☐ The circle below is about the same size as a penny. If this circle was a hole, you wouldn't have any trouble slipping a dime through it. It is possible to pass a quarter through the hole without tearing the paper. How would you do this?

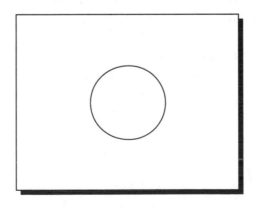

Problem Solving Style Tips

Problems are hurdles for the mental athlete. They represent the challenges that test your skill, and ultimately push you to be at your best. Rise to the challenge.

TIP ONE Before you start trying to solve a problem, make sure you know exactly what the problem is. Ask yourself what form the solution will take.

TIP TWO Organize the information you are given. Try to simplify the problem by restating it in your own words. Divide the problem into its component parts.

TIP THREE Do something to the information. Draw or sketch the facts, compare the problem with other problems, check your assumptions, and be ready to look in different directions. In short, perform a bit of intelligent groping in the dark.

*"The mere formulation of a problem
is often far more essential than its solution.
To raise new questions, new possibilities,
to regard old problems from a new angle
requires creative imagination
and marks real advances in science."*

ALBERT EINSTEIN, *physicist*

7

MENTAL PLAY

Fooling Around in Your Head

A Little Bent

Sit back for a few moments.
Look around the room you're now in.

Collect your mental energy
and, with all your might,

keep the furniture from laughing at you.

Fun and Frolic

*"Intelligence is the ability to see many points of view
without going completely bonkers."*

DOUGLAS ADAMS, *author*

When film-maker Steven Spielberg was designing the look for his movie *Close Encounters of the Third Kind*, he needed a break. Late one night, he decided to drive to the top of Hollywood Hill and gaze across the lights of Los Angeles. For no good reason, he did a head stand on the roof of his car and looked around. What he saw triggered an inspiration for the design of the mother ship in the movie, which was cityscape on the underside of the spacecraft.

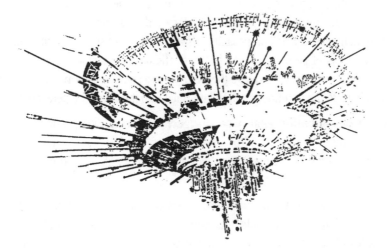

Very often, new ideas appear when you're relaxing or toying about. Make it a habit to take an occasional break from work. When you're least expecting it, your best ideas will magically appear. Here are some exercises to encourage an offbeat way of thinking.

Bouncing Off the Wall

A fun way to get your mind moving in new directions is to ask yourself, "What if..." followed by a hypothetical situation or condition.

- **What if the sky had always been dark?** How would human beings have evolved? Would our sense of hearing have become more developed than our sense of sight? We we receive sonar images instead of visual images? If so, what would these images be like? Could we *see* air cavities in our throat, lungs, and stomach? Would the reflective surface of our clothing become more important to fashion than color? If sonar was our main sense, what would books be like? How would people eventually discover clouds, or atoms, or red?

- **What if time ran backwards?** Would our usual sense of good and bad be inverted? Firemen would be bad guys because they would approach a building, flood it with water, watch it burst into flames, and then drive away as fast as they could. An arsonist, on the other hand, would be a good guy because when the firemen left, he would approach the fire, watch it get smaller until it soaked into the pools of gasoline he would pour. The arsonist would contain the fire into the head of a match and take the gasoline to a sucking station, which would send it back to the unrefining station and put it into the ground where it belongs.

- **What if our attention span were longer?** If we could keep our awareness steady for a few hours instead of a few minutes, would television commercials be longer? Would *War and Peace* seem too short? Would we be more or less willing to converse with boring people?

- **What if we lived to be six hundred years old?** When would people decide to have children? How would this affect insurance premiums and mortuary tables? Would people be more or less willing to skydive and to take other risks? Would we be more interested in long-term financial planning?

- **What if there were three sexes instead of two?** How would families change? What would pick-up bars be like? Would there be more or less gender discrimination? What would the third sex wear? Which sex would be the most powerful?

- **What if humor were outlawed?** Would the offence for telling a bad joke be worse than telling a good joke? Would people lose their sense of humor? Would there be underground schools to teach what it means to be funny? Would laughter seem like a pathological state?

- **What if people could feel the pain they afflicted on other people and animals?** Would there still be wars? Would there still be horse-racing? Would most of humanity become vegetarians? Would fishing become a thing of the past? In what other ways could people vent their aggression?

Another way to get your mind traveling in different directions is to visualize what it would be like to experience something you cannot perceive directly.

- What is it like to be a member of the opposite sex?
- What is it like to be a butterfly dreaming you're a philosopher?
- What is it like to be a brain cell?
- What is it like to be a protein molecule in a strand of DNA?
- What is it like to be a proton?
- What is it like to be the last sentence of a great novel?
- What is it like to be the Indian Ocean?
- What is it like to be a pigeon that roosts in Times Square?
- What is it like to be a hurricane?
- What is it like to be a cigarette?
- What is it like to have loved and lost?
- What is it like to be a snake eating its tail?
- What is it like to be a hallucinogenic drug?
- What is it like to be a wide open space?
- What is it like to be eternity?
- What is it like to be God?

Mental Manipulation

Below is a figure called the Necker cube. It was used by a nineteenth-century psychologist to illustrate how our perceptions are affected by our thoughts. When you look at the figure, you see the cube facing in one of two ways: the left-most square can appear be either at the front of the cube or at the back of the cube. Continue to look and the cube shifts — the cube seems to be facing in the other direction. Look at it for a while and notice how the perspective flips back and forth.

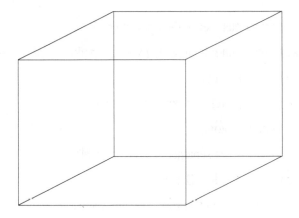

Now study the Necker cube and see if you can make the image shift at will. Make the cube appear in one perspective, then the other. Flip back and forth. Once you're feeling proficient, see the figure as a flat two-dimensional drawing, then as a cube seen from the left, then as a cube seen from the right. If you're really feeling ambitious, try to see the cube from the left and right simultaneously. Entertain the notion that you can perceive both views at the same time. What do you need to shift in your mind to do this?

The Möbius Strip

A mathematician confided
That a Möbius strip is one-sided.
You'll get quite a laugh
If you cut it in half,
For it stays in one piece when divided.

ANONYMOUS

One of the most elegant mathematical figures is the Möbius strip, a band with a half twist. Take a strip of paper, and join the ends with a half twist. If you cut the strip in half along the middle, you get a single band double the length with a full twist.

An even more astonishing trick is to divide a Möbius strip into three. Start the cut one third the width of the band, and cut around the figure twice. The result is that you get two linked bands, one of which is the same length as the original Möbius band, but one third the width, and another band which is double the length of the original with two full twists.

Some inventions actually use the Möbius strip. A Möbius film strip records sounds on both sides. Möbius strip tape recorders run the twisted tapes for twice as long. A Möbius strip conveyor belt is designed to wear equally on both sides, as is a twisted abrasive belt.

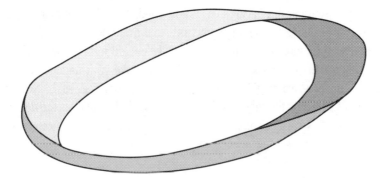

The Incomparable Mulla Nasrudin

Everyone should have a personal source of the bizarre, the paradoxical, the illogical. Throughout the Middle East, one popular source is Mulla Nasrudin stories — short anecdotes about the illogical but wise Mulla. At first, the stories don't seem to make a whole lot of sense, but if you roll them around in your mind for a little while, you may pick up some subtle meanings.

- Mulla Nasrudin was carrying home some liver which he had just bought. In the other hand he had a recipe for liver pie a friend had given him. Suddenly a buzzard swooped down and carried off the liver. "You fool!" shouted Mulla Nasrudin. "The meat is all very well — but I still have the recipe!"

- Mulla Nasrudin was walking through the streets at midnight. The watchman asked, "What are you doing out so late, Mulla?" Nasrudin replied, "My sleep has disappeared and I am looking for it."

- Nasrudin's wife burst into the Mulla's room and cried, "Mulla, your donkey has disappeared!" The Mulla calmly looked up and replied, "Thank goodness I wasn't on it at the time, otherwise I would have disappeared too!"

- A monk said to Mulla Nasrudin, "I am so detached that I never think of myself, only of others." Nasrudin responded, "I am so objective that I can look at myself as if I *were* another person; so I can afford to think of myself."

- When the village magistrate went out of town, Mulla Nasrudin acted as the judge. One day, a stranger ran into the court and cried, "I have been ambushed and robbed! Just outside this village someone stole my robe, my sword, even my boots. The thief must be from this village. I demand justice." Mulla Nasrudin looked at the man and asked, "Did the thief not take your undershirt, which I see you are now wearing?" The stranger, looking confused, said, "No, he did not." Mulla Nasrudin thought for a moment and said, "In that case, the thief could not be from this village. Things are done thoroughly here. I cannot investigate your case."

Mental Teasers

Zen Buddhism thrives on logical inconsistencies to show the limitations of rational thinking (and of other kinds of thought as well). Zen has been characterized as a spiritual teaching without scriptures, that is beyond words, that points to the mind-essence of man, seeing directly into his nature and leading to enlightenment. Traditionally, one of Zen's teaching tools is the *koan*, an illogical story that puzzles the student and makes him confront a different level of reality.

- A man traveling across a field encountered a hungry tiger. The man ran to a precipice, caught hold of a vine, and lowered himself over the edge, just out of the tiger's reach. Trembling, he looked down far below and saw another tiger waiting to eat him. Only the vine offered any protection. Two mice, one white and one black, little by little began to gnaw away at the vine. As the man looked at the vine fraying away, he noticed a luscious strawberry near him. Grasping the vine with one hand, he plucked the strawberry with the other. How sweet it tasted!

- Two Zen students were looking at a flag rattling in the wind and discussing what was happening. "It is the flag that is moving," said one. "No. You are wrong," said the other. "It is the wind that is moving." A master who overheard the quarrel approached the students and said, "You are both incorrect. It is your mind that is moving."

- A famous Zen master walked into town carrying a linen bag on his shoulder. Two Zen students, recognizing the simplicity of the man, approached him and asked him some questions. "What is the significance of Zen?" asked one. The man immediately plopped his sack down on the ground in silent answer. "Then," the other asked, "what is the actualization of Zen?" At once, the stranger swung the sack over his shoulder and continued on his way.

Interesting Thoughts

"What is mind?
No matter.
What is matter?
Never mind."

T. H. KEY, *19th-century scholar*

Keep a record of ideas that nudge you out of your usual way of thinking. Ideas about everyday things — they can be jokes, aphorisms, sayings, paradoxes, or bizarre statements — will come from many different places. Look for thoughts that provide insight, even if you're not exactly sure what the insight means.

Science fiction writer Stanislav Lec has an extraordinary knack for making the obvious sound profound. These are some of his observations.

- Sometimes you have to be silent to be heard.
- The exit is usually where the entrance was.
- On every summit you are on the brink of an abyss.
- The first condition of immortality is death.
- Dark windows are often a very clear proof.
- Sometimes the bell swings the bellman.
- Beyond each corner, new directions lie in wait.
- The moment of recognizing your own lack of talent is a flash of genius.
- A tired exclamation mark is a question mark.
- An empty envelope that is sealed contains a secret.
- Even his ignorance is encyclopedic.
- Think before you think!

Simply Amazing

Find the way from *Start* to *Finish*. Follow the direction that each arrow points until you get to another arrow. If you arrive at a double arrow, you can choose to go in either direction.

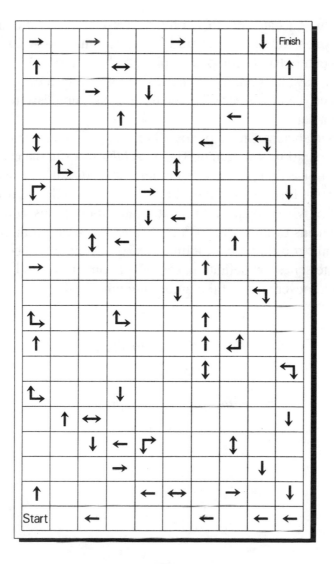

133

Vicious Circles

A paradox is a statement that contradicts itself. One of the first and probably best examples of a paradox was documented by Eubulides, a sixth-century B.C. Greek philosopher. In this paradox, Epiminides, a Cretan, says all Cretans are liars. If he is telling the truth, he is lying. If he is lying, he is telling the truth. Is Epiminides a truth teller or is he a liar?

Another ancient Greek philosopher, Zeno of Elea, produced a series of paradoxes of infinity. The most famous example is the race course or bisection paradox. If a rabbit runs a distance of one mile, then it must first run half the distance or half a mile, then it must run half of what remains or a quarter mile, then it must run half of what remains or an eighth of a mile, and so on ad infinitum. The rabbit must run through an infinite series of finite distances. Since an infinite series, by definition, can never come to an end, the rabbit should never arrive at the end of the mile.

"What Plato is about to say is false."

SOCRATES

"Socrates has just spoken truly."

PLATO

134

Russell's Paradox

Philosopher Bertrand Russell was renowned for his dislike of paradoxes and he spent a great deal of thought trying to explain them away. Along the way he formulated a paradox which now bears his name. It goes something like this:

A class is a collection of things. For example, a teaspoon is a member of the class-of-teaspoons. The class-of-teaspoons, itself, is not a teaspoon. and is therefore not a member of itself. The class-of-all-classes is a class. So the class-of-all-classes is a member of itself. The class-of-teaspoons is a member of the class-which-is-not-members-of-itself. Thus there is a class-of-classes-which-are-not-members-of-themselves. Is the class-of-classes-which-are-not-members-of-themselves a member of itself? If it is a member of itself, then it does not have the defining property and is not a member of itself. And, if it is not a member of itself, it does have the defining property and has to be a member of itself. Each alternative leads to its opposite.

Russell lost many nights' sleep over this paradox. He wrote, "Every morning, I would sit down before a blank sheet of paper. Throughout the day, with a brief interval for lunch, I would stare at the blank sheet. It seemed likely that the whole of my life might be consumed in looking at the blank sheet of paper."

Can you form a clear image of Russell's paradox? Can you draw a picture of it? Can you imagine why this paradox would be important?

Sprouts

Mathematician John Conway developed a fascinating and easy-to-learn game called Sprouts. The playing surface is a group of sixteen dots arranged in a four-by-four grid. Two players alternate moves by connecting any two dots together with a straight or curved line. A new dot is placed on this line. Lines cannot intersect and a dot may have at most three lines connected to it. The aim of the game is to have the last move.

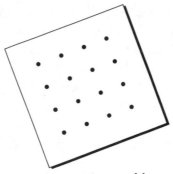

The game produces many interesting and beautiful patterns. While some people draw short stark lines, others draw long flowing lines. Some people reach far out with their moves, others play close in. As a result, a sprout game can become an aesthetic endeavor as well a battlefield for the intellect.

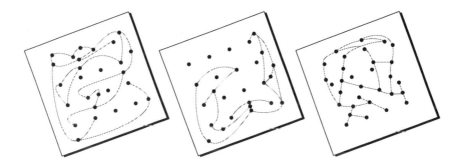

Go

Go is an ancient game, developed in China over four thousand years ago. The rules of Go are simple. On a nineteen by nineteen square playing surface, two players alternate placing stones on the intersection of the lines. The stones are not moved unless captured. The aim is to enclose as much area as possible as well as to capture the opponent's stones. If a group of white stones is completely surrounded by black stones, the white stones are considered dead and are removed from the board.

Go is easy to learn but difficult to master. To do well in Go, it's best to think in visual patterns. Go masters speak of game positions in terms of living groups, dead groups, initiative, breathing spaces, strong and weak shapes, lines of communication, and armies that extend fingers into enemy territories.

Go engages many types of thinking skills from visual strategy, in the opening stages, to logical tactics, in the middle and final stages of the game. In fact, the game was played by seventeenth-century samurai in Japan to teach warfare. Today, in Japan, China, and Korea, the game is played for its aesthetics and its ability to teach initiative, sacrifice, and risk taking.

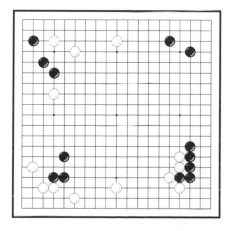

Mental Play Style Tips

The mental athlete knows that humor lifts the spirit and refreshes the mind. Make it a habit to play the part of the jester.

TIP ONE   Periodically, look at things upside down. Stand up, turn around, bend over, and look between your knees. Not only do you get a new perspective on things, you let a healthy dose of blood flow to your brain.

TIP TWO   Search for ideas that are bizarre, unusual, shocking, off the wall, preposterous, deep, and fun. Keep a stock of these in your mind and pull them out when you need them.

TIP THREE   If you're feeling as if you are under a lot of pressure, if you're taking things a bit too seriously, or if your thinking is stuck in a rut, then take a break.

"There are some things that are so serious
that you have to laugh at them."

NIELS BOHR, *physicist*

S T A T I O N

8

REACHING BACK

Memory Muscles

Looking Backwards

*Spend a few moments recalling the following items.
As you do, pay attention
to the quality of your recollections.*

*How clear and vivid are they?
How fast do they appear?*

*Take time
for these memories
to form fully in your mind.*

*Recall the last time you went swimming.
Recall the name of your first grade teacher.
Recall what the second chapter of this book is about.
Recall the name of the capital of Canada.
Recall the smell of garlic.
Recall why the sky is blue.
Recall what you did on your twelfth birthday.
Recall a time when your memory served you well.
Recall the square root of sixteen.
Recall what clothes you wore yesterday.
Recall the shape of an oak leaf.
Recall a very early memory.
Recall when you began this exercise.*

The Passing of Time

"Memory is the treasury and guardian of all things."

CICERO, *1st-century B.C. orator*

One of the earliest memories of child psychologist Jean Piaget was of a kidnapping attempt when he was a toddler. Piaget vividly remembers being attacked by two men as was being carried to a car by his nanny. The nanny fought the men off and she and the child safely escaped into the car. As a reward, the family gave the nanny a gold watch. Twenty-five years later, the nanny confessed that she had made up the whole story to get a raise. Yet Piaget could still recall the event vividly, even though it never took place.

Cognitive psychologists say that memories are not duplicate impressions of earlier events, but *reconstructions* of earlier events. Because young Piaget heard the story of the kidnapping so many times, the pictures in his mind became real, giving the impression that they actually took place.

If you recall the last time you went swimming, you'll likely see a picture of yourself floating in the water. But think about this for a moment, and you'll realize that is not the way you actually experience swimming. When you swim, you experience the sensation of floating, the wetness of the water, the splash over your face and eyes. Yet, in your imagination, you experience it from another point of view. You remember what you tell yourself has happened.

Why do we remember some things, sometimes, and forget the same things at other times? Why do some memories seem to fade into the distance with time, losing their original flavor, and others remain as vivid as if they occurred yesterday? Why can we recall the characters in a novel we read five years ago, but not the name of someone we met yesterday?

The Anatomy of Memory

Psychologists tell us there are three distinct stages of memory. To demonstrate these stages, glance at the following series of letters.

O I C U R M T

Glance at the letters again. This time close your eyes almost immediately. You'll notice there is an afterimage of the shapes. Even though the afterimage lasts only for an instant — a second at most — it plays a role in the memory process. Immediately after you see, hear, or feel something, there is a fragile, brief lingering of the perception. Visual images are the shortest in duration, auditory images last slightly longer, and sensory images can last several seconds. This lingering is called sensory register memory. Sensory register memory is primarily involved in smoothing out the continuity in our perceptions. The world does not suddenly go blank when we blink.

Now look at the letters again. This time, hold the letters in your mind first by repeating the sounds and then by picturing the shape of the letters. When you do this, you are using what cognitive psychologists call short-term memory. Short-term memory is in the forefront of your consciousness. Unlike sensory register memory, which is engaged automatically and involuntarily, short-term memory can be focused with an act of will.

Long-term memory, by far the largest component of your memory system, is your permanent storehouse of information. It contains concrete memories, such as the first time you rode a bicycle, your knowledge of where you live, your sense of the English language, as well as more abstract memories such as your personal values. Without long-term memory, we would be constantly living on the thin edge of the present, as if we were perpetually waking up from a dream.

These are the three stages of memory.

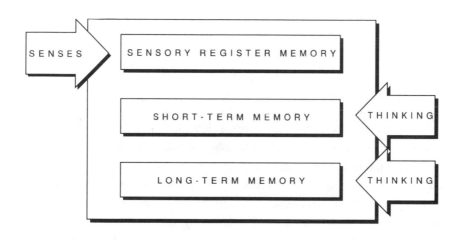

For something to be entered into memory, it must pass through all three stages. Though some experiences are transferred automatically from short-term memory to long-term memory — events that are particularly meaningful or have a strong emotional impact on us — most experiences don't. Improving your memory involves learning how to pass information from one stage to another.

To recall words, names, numbers, lists of things to do and buy, information must be *encoded* in short-term memory. There are several ways of doing this. The more deeply you process or think about something, the more ways you have to index a piece of information, the more associations you have with it, then the easier it is to remember. By finding patterns, hooks, images, associations, and meanings, memories become richer and more accessible.

> *"It's not how you take it out,*
> *it's how you put it in."*
>
> LINDA PERIGO MOORE, *author*

Figure Out What's Important

"To expect a man to retain everything...
is like expecting him to carry about in his body
everything he has ever eaten."

ARTHUR SCHOPENHAUER, *philosopher*

People tend to remember what's relevant to them. That's why a fashion designer can easily remember shades of color, why an international banker can recite exchange rates, and why an astronomer can instantly recognize star patterns. In each case, the subject holds meaning for them. Because we remember what's important to us, we can enhance our memory by deciding whether something is, in fact, important. If it is important, we naturally pay more attention. We give ourselves the motivation to recall.

It's true that we can't always decide beforehand how important something will be, or whether something will be worth remembering. But at times we can judge whether we will need some information in the future. Here are three ways to help you decide:

❏ **ASK YOURSELF** "What would happen if I don't remember this? Would it really make any difference?" If it wouldn't, then don't bother remembering it.

❏ **ASK YOURSELF** "How soon do I need to use this information?" If you need it sooner, then it has a greater impact.

❏ **ASK YOURSELF** "Does anything else depend on remembering this information?" A subject's real importance is measured by its value relative to other subjects.

In the process of deciding whether something is worth remembering, you determine its importance. If you think something makes a real difference to you and other people, if it carries immediate impact, it takes the top of the list. Giving a piece of information high priority automatically encourages you to pay more attention.

To keep your memory muscles in shape, it's helpful to develop the habit of paying close attention. For example, next time you put down your house keys or car keys, consciously experience setting them down, rather than just dropping them off thoughtlessly. If you place them on a table, look around you. See the surface on which the keys are resting. Ask yourself, is it dark or light? Is it smooth or rough? Is it high or low or somewhere in between? Look at the keys. Are they lying fanned out or piled on top of each other? Is the key ring lying flat or at an angle? Picture the keys in your mind. Touch the keys and the surface. What do they feel like? Engage all your senses. Can you smell anything? Is the air warm or cool? You never know what cue may trigger your recollection. In the beginning, it may take several seconds to build up your impression. Later, it takes an instant.

Similarly, during the next conversation you have, pay close attention. The main reason we forget what we hear is that we are not really listening. Either we let our minds wander, or we focus on what we want to say ourselves. Being a good listener involves focusing our attention on what the other person is actually saying. Remember the saying, "The best conversations are those in which you focus on the meanings, not the words."

Remembering a person's name involves the same skills in remembering where you put things, and remembering what people say. When you meet someone whose name you want to remember, pay attention with all your senses. Listen to the sound of their name. Does it have a hard sound, like Dick or Greg? Or a soft, gentle sound, such as Michelle? Is the sound long and stretched out, like Jennifer? Or short and compact, like Tim? When you say the person's name — "Hello, Frank", "Kornell, is that your name?" — listen to the sound roll off your tongue.

Ask yourself if the name seems to fit the person. To decide, look closely. Does he or she appear strong and sturdy, weak and frail, hard or soft? What kind of body type does the person have? Notice the person's hair, eyes, and skin color. Listen to the sound of the person's voice, as well as the name. If you shake hands, feel the grip and the texture of the skin. Can you smell a perfume or after-shave? Make a connection between your impression of the person and the sound of his or her name. Tim is the tall, bald man with the firm handshake, and the soft, lyrical voice. Chris is the short, dark-haired woman with round features.

CHAD?

GRIMSBY?   ENGELBERT?

CORNELIUS?   IRVING?

KENNETH?   BARTHOLOMEW?

ISAAC?   ETHELNED?

HANK?   GARTH?

CHUCK?   ALARIC?

SAM?   MURRAY?

ERWIN?   JO-JO?

BILL?   FRANCIS?

GARFIELD?   ARNOLD?

TIP If you want to remember someone, pay attention to them.

Hooks

"The horror of that moment," the King went on,
"I shall never, never forget!"
"You will, though," the Queen said,
"if you don't make a memorandum of it."

LEWIS CARROLL, *mathematician and author*

Can you recall the sequence of letters five pages ago? If you originally tried to memorize the sequence by repeating the letters by rote,
you probably can't recall them very well. But if you tried to find a
hook, a pattern, another meaning, you may have recognized that
the sounds of the letters, O I C U R M T, form the phrase, "Oh, I see
you are empty." Once you know this, the letters become easy to
recall.

In 1968, Alexander Luria, a cognitive psychologist, studied the
remarkable memory of a Russian newspaper reporter referred to as
S. Through imagery and association, S. could effortlessly remember
astonishingly long lists of digits and names. To encode a grocery
list, for example, he would imagine himself walking down Gorky
Street and picture each item at a specific place along the street. He
might place the eggs under a street lamp, a quart of milk in a foun-
tain, and a pound of butter in a tree. To recall the objects, he would
mentally walk down the street again, examine the landscape, and
recite the objects that he saw. Curiously, he only made mistakes
when he put an object in a hard-to-see place, such as the egg in
front of a white wall, or licorice in the shadow of a building.

This memory technique, used by early Greek orators, is called the
method of *loci*. First, you thoroughly memorize a familiar environment. It may be your home street, your way to work, your kitchen,
or your bedroom. Next you put the objects that you want to remember into specific locations in the environment. If you had a grocery
list of milk, eggs, artichokes, steak, wine, you would mentally place

these objects, one at a time, into the environment — put a container of milk at the foot of your bed, a carton of eggs on your dresser, a giant artichoke on your bedroom telephone, and so on. Try to make the objects large and conspicuous. The more bizarre the picture looks, the better you'll recall it.

A similar technique can help you remember a list of abstract ideas. Say you need to give a speech or presentation. Translate the abstract ideas into visual images of tangible objects. If you need to make a point about profit, visualize a business graph. If you need to mention product distribution, picture a truck or train. If you need to talk about a change in management attitude, envision a manager upside-down. When you plan out your presentation, arrange these visual images in a sequence — a graph pasted in the front of a truck which is being driven by the upside-down manager. When you need to recall the series of ideas, recall the objects.

Remembering Exercises

❑ **Can you recall ...**

What were you thinking about five minutes ago?
What were you thinking about an hour ago?
What were you doing yesterday at this time?
What you had for breakfast for the last three days?
What you wore last weekend?

❑ **Settling Back**

What are some of your earliest memories?
What are some of your most vivid memories?
What are some of your dimmest memories?
What kinds of things do you have a good memory for?
What kinds of things do you have a poor memory for?

❑ **Short-term Attention**

Sit across from another person, select a sentence at random from a book, and read it to yourself. Look the other person in the eyes, and recite the sentence exactly. The other person then repeats what you said. If he or she makes a mistake, repeat the sentence. When the person repeats the sentence correctly, move on to another. Choose short sentences at first and move to longer, more difficult sentences. After a while switch roles.

❑ **Dinner Picture**

Next time you have dinner at a restaurant, take a mental snapshot of what your place setting looks like. Do this by visualizing an imaginary connection between one object on the table and another. For example, the salt shaker falls onto the dinner plate, which in turn spins, causing the fork and knife on the plate to strike the wine glass, causing the vase to tumble making the six flowers land on the water glass, which in turn ...

❑ **Reflections**

Recall something you told yourself that you would never forget.

❏ Visual Memory

Look at the following figures for a minute, then draw them as accurately as you can from memory.

❏ Daily Recall — Outside

When you go to bed at night, before drifting off to sleep, visualize the day's events. Picture what happened to you, starting from the moment you awoke. Visualize what your life would look like if a hidden camera were following you around all day.

❏ Daily Recall — Inside

Visualize the day's events as you drift off to sleep. But instead of visualizing your life as seen through a hidden camera, visualize what you saw through your own eyes.

❏ Daily Recall — Sensory Recall

As you go to sleep, recall all that you heard throughout the day. Do the same for what you smelled, what you tasted, and what sensations you experienced.

❑ Remembering a Memory

Spend a few minutes examining your shoes. Realize that as you are doing this, you are creating memories. After ten minutes, recall the memory of examining the shoe. How does it feel? What differences are there? Five minutes later, recall the memory of examining the shoe. Has it changed? Recall the memory of the memory of examining the shoe you had five minutes earlier.

❑ Memory for Will Power

Choose a simple action such as tying your shoes, scratching your nose, or stretching your legs, and decide that you will perform this action at a certain time later today. For example, decide that at 6:00 P.M., you will clean your glasses. What might make you forget to do this?

❑ Peg Words

Memorize this list of peg words.

One is a bun	Six is sticks
Two is a shoe	Seven is heaven
Three is a tree	Eight is a gate
Four is a door	Nine is a line
Five is a hive	Ten is a hen

Using this list, it's possible to recall ten items fairly easily. Imagine that you needed ten tools from a hardware store. Mentally place the first tool — say a hammer — in the bun, the second tool — a screwdriver — in the shoe, and so on, until you have associated each object you need to buy with an object in the peg word list. Then recall the hardware tools by recalling the list. Practice by memorizing this list and recalling it tomorrow at lunch.

hammer	nails
screwdriver	washers
saw	electrical wire
level	light bulbs
putty knife	drill bit

Next time you have a shopping list that has about ten items, use the peg word method and save paper.

The Present Time

*"Can you remember
what remembering was like last Tuesday?"*

LUDWIG WITTGENSTEIN, *philosopher*

Memory gives us a sense of personal continuity and direction. Our everyday thoughts include plans, expectations, and recollections, which frame a sense of context.

We can often learn more about memory by studying people whose memory is not effective. A classic example of a memory-deficient person is H.M. After brain surgery to cure his epilepsy, H.M. could no longer remember anything new. His early memories — of his family, his friends, and his home — were intact, but he could not recognize any of the nurses, nor could he learn his way to the bathroom.

H.M. described his experience as feeling as if he were continually awakening from a dream. If his attention was distracted, he would have no idea where he was, what he was doing, or what day it was.

It's worthwhile exploring how our thoughts create a sense of context for our mental life. Memories seem to be located behind us. We look from the here-and-now back to the past and get a sense of progression of time. Expectations and plans seem to be located ahead of the present. We mentally gaze forward, as if we were traveling to this destination.

But these impressions are just a way of indexing our experience. All our experiences take place in the present moment. Each and every thought — those that refer to future events and those that refer to past events — is experienced in real time.

To think more about this puzzling aspect of thought and time, try the following exercise.

Past – Present – Future Thoughts

Spend a few minutes relaxing
and letting your attention loosen up.

Allow your breathing to slow down
and to become smooth and even.

When you feel refreshed,
pay attention to your thoughts.

Notice that some thoughts
refer to the past,
some to the present,
and some to the future.

Spend a while paying attention
to this past-present-future quality.

Notice how the direction of your thoughts,
the sense of looking back and looking forward,
creates an impression that you are "going somewhere,"
and that you are moving from the past into the future.

As an experiment, try adjusting this quality.
Choose a specific thought,
say an image of an upcoming event or of something past,
and reverse the sense of direction.

For example, visualize yourself as being in the future
looking back towards the present,
or alternatively, see yourself in the past
looking forward into the present.

How does your sense of the present change with this exercise?

Memory Style Tips

Our memory is not only what we remember with, but what we forget with. Keep your memory muscles limber by keeping your attention in shape.

TIP ONE Pay attention only to what you want to remember. Don't try to retain everything; select only the items you want to recall and don't bother with the rest.

TIP TWO Organize material you want to recollect into a framework. Learn the material as a whole rather than in parts. Focus on meaning and context.

TIP THREE Use cues to help you learn the material. Visualize, associate, put your ideas into a story. This will assist in getting the information more firmly embedded in long-term memory. Develop your own techniques to help you remember better.

TIP FOUR Keep your memory alive by periodically exercising your recall muscles. Every once in a while, ask yourself what you want and need to remember from the past. What lessons have you learned yesterday and today that you want to remember tomorrow?

"What we remember can be changed.
What we forget we are always."
RICHARD SHELTON, *author*

9

MENTAL FLEXIBILITY

Analysis and Synthesis

Switch Modes

*Look at the illustration below.
Try to discover how the complex pattern
is created out of the simpler shapes.*

Mental Jumping

*"All men see the same objects,
but do not equally understand them.
Intelligence is the tongue
that discerns and tastes them."*

THOMAS TRAHERNE, *17th-century poet*

To make sense out of the blur of sights, sounds, textures, aromas, and tastes that make up life, we constantly use two mental abilities, analysis and synthesis. Analysis is the ability to divide big ideas into little ideas. When you analyze a relationship between two people, the risk in a business venture, or the progress of a baseball game, you examine the whole by dividing it into smaller conceptual components. Synthesis, the flip side of analysis, is the ability to combine many little bits of information into a few general principles. When you synthesize what you know about the economics of the country, how gravity works, or the direction your life is going, you group fragments of information and detect patterns, relationships, and underlying causes that explain the big picture.

Consider the pattern on the facing page. What do you need to do to comprehend it? First divide it into components by focusing in on only one type of shape — first stars, then triangles, then hexagons, and then pentagons. Notice how each of these shapes fits together with the others. Choose a line segment, follow its path through the pattern, and see how it establishes borders and outlines. Then, think of some general rules which underlie the pattern. Search for a plan that puts the lines and component shapes into order. After you're done, ask yourself if you could reconstruct the pattern.

Life can be spectacularly complex and disordered. To make sense of it, we put things into categories. Mentally, we parenthesize our experience.

Mental Breakdown

EXERCISE Describe what you look like.

Analysis is the process of making distinctions. To describe your appearance, you have to mentally divide your body into parts and then characterize each of these parts. You could start by describing some general features, such as your build and the overall proportions of your body, and then focus on details, such as the color of your hair, skin, and eyes, the size of your hands, the roundness of your face, the height of your forehead, the shape of your nose, the fullness of your lips, the curve of your chin. Without an ordered division, your description will be a jumbled arrangement of unrelated features.

Curiously, if you asked five people to describe your appearance, you would likely get five somewhat different descriptions. Everyone has his own way of interpreting what he sees and his own standard of comparison. One person sees a nose as large, another as hooked, another as crooked. This is why eyewitness identification is often contradictory.

Many skills — cattle-judging, bird-watching, gourmet cooking, diamond-sorting, wine-tasting, to name but a few — are based on the art of classification. The educated mind knows what to look for and what to ignore. An untrained mind might look at a tree and say "tree." The discerning mind looking at the same tree would perceive the shape of the leaves, the colors in the bark, the proportions of the limbs, and recognize the species, age, and health of the tree.

A problem arises when we become lazy in our choice of categories. We may get into the habit of immediately classifying things into what we like and what we dislike, without ever going beyond our initial reactions. We may see things from a single point of view. We may develop fixed opinions and beliefs. We may become so enmeshed in our mental patterns that we forget it is our mind that creates our experience. Change our mind and we enhance our experience.

A good way to exercise your powers of analysis is to adopt new categories to pour in your observations. Look around the room and choose an object. It could be a chair, shoe, or wallet. Look at that object and filter what you see through the following five categories.

Origin

Material

History

Current Use

Future

For example, if you look at the origins of a chair, you might notice features that suggest how the chair was made. Change your mindset to see the materials in the chair and you'll think about the kind of wood or fabric that makes it up. Consider the chair's history and you'll find clues to its past — scuff marks, paint flecks, dents, and pits. Look at the chair's current use and you might become philosophical and think about the countless purposes a chair can have, besides acting as a seat. Imagine what may lie in the future for the chair and you search for indications in the environment about what is going to happen to it.

With practice, you will see that even the most familiar objects — ironing boards, dusty old running shoes, pavement — become fascinating. The categories you establish guide your observations, and your observations stimulate you to find new categories in which to channel what you perceive.

Discovering Patterns

"To know truly is to know by causes."

FRANCIS BACON, *17th-century philosopher and scientist*

EXERCISE What is the next letter in this sequence:

An important part of our intelligence is our ability to find patterns. Gregor Mendel found a statistical relationship between dominant and recessive traits in offspring and produced the genetic theory. Charles Darwin saw patterns of physiology and behavior that led him to formulate the theory of evolution. George Hubble found a pattern between the distance of galaxies and the rate at which they were receding, and concluded that the universe is expanding. Dmitri Mendeleev discovered a relationship between chemical elements and formulated the periodic table that helped develop the atomic theory.

To solve the letter sequence problem, you need to discover a quality that the letters share. You may look at the letters' place in the alphabet for a clue. You may examine the shape of the letters. You might think that the letters represent other sequences. With this in mind, the pattern jumps out at you: the sequence is made of the first letters of the words one, two, three, four, and so on. The next letter in the series is the eighth in the series, and therefore is the letter E.

EXERCISE Look at the following shapes and ask yourself which is different from the rest.

If you thought the circle was different, then you would be correct, since it is the only shape that has no straight lines. However, if you thought that the square doesn't belong, then you would also be correct since the square is the only shape that has four right angles. If you thought that the triangle is different, you would be correct too, because it is the single object that is not symmetrical. The pie wedge is the only shape that contains straight lines and curved lines. And the final shape is the only one with a gap in it. In short, each shape is different from the others. In this sense, they all share a quality, which must mean that they are all similar.

A large part of our intelligence is our ability to recognize patterns. This pattern recognition ability enables us to recognize faces, to know when we are listening to Wagner, to recognize a friend's walk from a distance. Our tendency to formulate opinions and to make generalizations is also an extension of our urge to make the world a pattern.

TIP The right answer depends on what you are looking for.

Analysis Exercises

❏ **Division**

In how many ways can you divide a square into four identical parts? Aim for ten different ways.

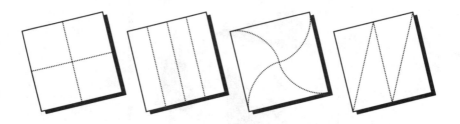

❏ **Critic**

If you were a critic, what criteria would you use to judge the quality of a record? Consider the relevant factors: the quality of playing and production, the emotional tone, the lyrics, the musical inventiveness. What criteria would you use to judge the following?

movies	television commercials
modern art	white wine
fashion	potato chips
cars	emeralds

❏ **Selective Attention**

Next time you listen to a symphony, listen to one particular instrument. Hold your attention there for the entire duration of the piece.

Next time you're eating dinner, analyze what you're eating. Break down the food into the caloric count. Try to identify each component in the food, by taste, texture, smell, and sight. Think of the origin of the food, and what has been involved to get it to your dish.

In how many ways can you analyze the human body? Consider the body's functional systems, structures, levels of organization, and development.

❏ Categories

Think of all the categories that you can for the following subjects:

Food: fruits, meats, proteins, carbohydrates, ...
Sciences: physics, paleontology, pseudoscience, ...
Music: rock, classical, musical instruments, marketing, ...
Law: courts, judges, police, verdicts, ...

❏ Sequencing

Another aspect of analysis is dividing up a large complex plan of action into a series of simpler actions. Write out the steps you need to perform to achieve the following projects. List possible alternatives if things don't go according to plan.

Buying and using a home computer
Making a fortune in real estate development
Training and keeping a seal
Traveling to Timbuktu
Becoming fluent in Farsi
Becoming a lighting designer for a rock band
Building an addition to your home
Making a hunting knife
Rewiring your house or apartment
Putting a new bathroom in your house
Organizing an immunization campaign in a Third-World country
Starting a newspaper
Designing a French restaurant

❏ Proportions

What proportion of your body space is solid, liquid, and gaseous?
What proportion of your day do you spend without seeing anyone else?
What proportion of your time do you spend eating, sleeping, resting, working, playing, daydreaming?
What proportion of days are clear, cloudy, rainy?
What proportion of your money do you spend on long-term goals? On entertainment? On gifts?
What proportion of time do you sit with your legs crossed?
What proportion of the books you read are novels? Science books? History? Drama?

Viewpoints

To help you be more flexible in your interpretations, practice considering situations from several alternative points of view. Imagine the following scenarios, and put yourself in other people's shoes.

❑ Your teenage son has just told you that he has a drug problem. Consider the situation from the viewpoint of the son. The parents. The boy's best friend. His teacher. His younger sister. His drug supplier.

❑ A single mother who shop lifts in order to make ends meet is caught trying to steal a diamond necklace. Consider the situation from the viewpoint of the mother. Of the police officer. Of the department store owner. Of the mother's grade-school son. What would the situation be like if she were caught stealing a loaf of bread? Pick-pocketing a wallet?

❑ A union goes on strike. Consider the situation from the viewpoint of the union leader. Of lower level management. Of the owner of the company. Of the workers who cross the picket line.

❑ You see a group of Hari Krishnas singing and dancing on your town's main street. How do feel about what you see? What would a junior member feel about it from the inside? How would the parent of one of the new members feel? What about the organizer?

❑ You see someone toss a candy-bar wrapper on a sidewalk. Consider the situation from the point of view of the litterer, of a street cleaner. (If people didn't litter, would the cleaner be out of a job?) Of a store owner. Of the candy-bar manufacturer.

❑ You see an elderly person struggling to get into a bus. Look at the situation from the point of view of the bus driver. Of the elderly person. Of a young person in a rush who is behind the elderly person.

Guesstimation

Another facet of analysis is the ability to guess how large or small, fast or slow something is. In everyday situations, our information is based on estimates. Exercise your ability to make educated guesses. Try these out:

❑ **Time**
How long does it take to wash dishes?
How long does it take to fill your car's tank with gas?
How long does it take to read this book?
How long does it take to paint your living room?
How long does it take to take a shower? To get ready from the moment you get out of bed to the moment you step out the front door?

❑ **Distance**
How far is it to your left index finger?
How far is it to the nearest nuclear power station?
How far do you travel in ten average steps?
How far is it to India?

❑ **Weight**
How heavy is a blade of grass?
How many eggs would equal the weight of one car?
What is the weight of air in the room you are now in?
How heavy is your head? Your left arm? Your legs?
How much water do you consume in a year? How much sugar? Salt?

❑ **Dimensions**
What are the dimensions of the rooms in your house or apartment?
What are the dimensions of your kitchen table, chairs, windows?
What are the vital statistics — height, weight, measurements, suit size — of your spouse? Choose three people and estimate their measurements.
How large is a blood cell, a carbon atom, the sun, a DNA strand? A computer chip, a virus, a dinosaur, a spiral galaxy?

Synthesis Exercises

❏ **The Truth of the Matter**
Consider the following statements. First, assume they are true, and give three plausible explanations for their validity. Then, assume they are false, and state three plausible explanations for their lack of validity.

More murders occur on Saturdays than on any other day of the week.
The position of the planets at your birth affects your destiny.
Human beings evolved from apes.
Television is chewing gum for the eyes.
All people are psychic under the right conditions.
Eating meat is bad for you.
Not expressing negative emotions is healthy.

❏ **Flipping Modes**
The usual tendency to label experience as good or bad causes us to respond to events in predictable ways. Try to reformulate some of these labels by thinking the opposite of what you might routinely think.

You have been fired. List five things that are good about it.
You have won a million dollars. List five things that are bad about it.
You have just missed an important deadline. What good can come of it?
You have won admiration for a radio appearance. What's bad about it?

❏ **Determine the next number or letter in the following sequences:**
A E F H I K L M ... ?
J30 J31 A31 S30 O31 N30 ... ?
98 34 14 10 ... ?
3 1 4 1 5 9 2 6 5 3 ... ?
Q W E R T ... ?

❑ **Since synthesis involves putting together bits of information, a good way to enhance your skill at it is to practice holding as much in your head as you can. Try solving the following problems by sorting out all the relationships between the facts.**

❑ Huey and Dewey are the same age. Huey is older than Lewey, who in turn is older than Tom. Dick is older than Tom, but is younger than Huey and Lewey. Dewey is younger than Harry. What is the order of age of the six people?

❑ Tom is twice as old as Dick will be when Harry is as old as Tom is now.
Who is the oldest?
Who is the youngest?
Who is in between?

❑ Tom, Dick, and Harry are tied in the welterweight Mental Olympics. To break the tie, they are blindfolded and told that either a red or white hat will be placed on their heads. If they see someone else wearing a red hat, they must raise a hand. When the blindfolds are removed, all three men raise their hands. Finally, after several minutes, Harry stands up and says, "I am wearing a red hat." How did he figure it out?

❑ If a hen and a half lays an egg and a half, in a day and a half, how many and a half hens, who lay better by half, will lay half a score and a half in a week and a half?

❑ **Alternative Explanations**
A good way to keep your synthesis muscles in shape is to invent a number of hypotheses of what's going on. For example, say that you have not received any telephone calls this week. You may hypothesize that you haven't called anyone else, that your friends are all sick, that it is a busy time of year in your industry. Consider these situations.

Someone at the office suddenly takes an active interest in you.
It seems unnaturally quiet in your neighborhood this morning.
Your legs are stiff this morning.
You have an unusually refreshing night's sleep.

Searching for Evidence

EXERCISE You are told that cards with light gray faces have circles on the other side. Before you are four cards, two face up and two face down. What is the minimum number of cards you must turn over to test whether the statement that all light gray cards have circles on their other side is true or false?

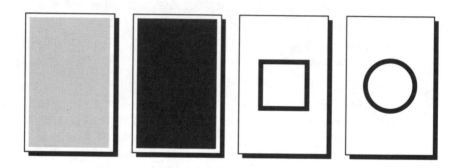

The correct answer is two. You must flip the one showing the light gray face, and the one showing the square. If the reverse side of the light gray card is a circle, the rule is confirmed. If it is something else the rule is false. But this does not give you all the evidence you need. You must also look on the opposite side of the square. If that side shows a light gray face, the proposed rule is false, because you have found a light gray card which does not have a circle on the other side. If it is dark gray, or anything else, the rule remains intact.

This simple card trick illustrates a tendency that many of us have when we search for evidence. Many people opt to turn over the cards that *confirm* the rule. They turn over the card with the light gray face, and leave it at that. Or they may also turn over the card with the circle. The circle card does not add to the evidence because it could show either a light gray or a dark gray surface with the rule being true. Dark gray cards may also have circles.

The tendency to look only for confirming evidence and neglect denying evidence occurs in many areas of life. If we belong to one political party we tend to see only the good things that party has done. Many people who believe in astrology, parapsychology, and psychic events tend to look at the hits and ignore the misses. Part of the reason is that we often want the world to work a certain way.

A good scientist neither believes nor disbelieves ideas — she tests them. She asks questions about a subject and then formulates a hypothesis. If she is working on the extinction of dinosaurs, she may formulate the hypothesis that the giant lizards were wiped out by a massive asteroid impact. Based on this hypothesis, she devises an experiment and predicts what the outcome will be. She could predict that there must be some evidence for an impact crater and then look for the crater, or she could predict there must be residue from the asteroid left in sedimentary rockes. If the observations match the prediction, the theory is on the right track. If the observation does not match the hypothesis, the theory needs adjustment.

A hypothesis can never be proven to be true. It can only be sup-ported or rejected. If an observation fits a hypothesis, then the hypothesis is reinforced. Each time the hypothesis is tested and fits the data, it becomes more useful as an explanation of how things work. If further research turns up observations that don't fit, the hypothesis must be rejected or modified. This modified hypothesis must, in turn, be tested. This is how our knowledge advances.

The scientific method can be applied to many things. Say you wanted to find out if red sunsets mean that bad weather is approach-ing. You should watch for red sunsets and keep track of the weather the next day.What would you do if you wanted to test the idea that unselfish actions are highly rewarding, or that mental exercise makes you more creative, or that telepathy is possible? Devise experiments to scrutinize these hypotheses. When you step back a pace from your opinions and are willing to see your experiment succeed or fail, you are following the scientific method.

TIP Scrutinize your beliefs. Ask yourself what evidence supports or rejects your position.

Analysis and Synthesis Style Tips

The abilities to analyze and synthesize are like complementary mental muscles. Analysis lets you break things apart and synthesis lets you put things together. Together they provide mental articulation and precise thought.

TIP ONE   Make it a habit to be aware of how you mentally divide your experience. What are the categories you use to order your observations?

TIP TWO   Recognize the patterns you use to order the world. Practice switching points of view. Put yourself in other people's shoes by seeing things through their eyes.

TIP THREE   Use the scientific method whenever possible. Don't believe things at face value, but observe carefully, test ideas, be willing to adjust your opinion based on the evidence you gather.

*"No amount of experimentation
can ever prove me right;
a single experiment can prove me wrong."*
ALBERT EINSTEIN, *physicist*

10

MENTAL BALANCE

Decision Making

Heads or Tails

You are walking on a beach
and notice an ornate bottle
half buried in the sand.

As you pick up and open the bottle,
a mist rises and materializes into a magical being:
a genie.

Unlike most genies,
this one doesn't give three wishes,
he grants three choices.

One:
You can have five years added to your life
under the condition that another person,
selected at random,
loses five years of his or her life.
Would you take the extra time?

Two:
You can have twenty thousand dollars,
if you agree to have a tattoo the size of a dollar bill.
Would you take the money?
If so, where would you place the tattoo,
and what design would you choose?

Three:
When you awake tomorrow morning,
you can have a single new skill or attribute.
What would you choose?

The Anatomy of Choice

"The strongest principle of growth lies in human choice."

GEORGE ELIOT, *19th-century novelist*

How many decisions do you make in a day? Ten? Twenty? A hundred? A thousand? You make action decisions every moment. The way you sit, the words you choose to use, the manner in which you respond to your friends and family — although lodged in habit and routine — are the products of choice. You make intellectual decisions less frequently. The clothes you buy, the television shows you watch, the cereals you eat, are choices that you make occasionally. Both types of decisions define your character. In the words of management consultant John Arnold, "What you choose is what you are."

Many choices, like those you confronted in the opening exercise, depend on factors other than logic or reason. The purpose of the exercise was to help you think about how you make difficult choices. Did you evaluate all possible alternatives? Did you consult your gut? Did you make a snap judgement?

To make wise choices, it's helpful to have a method for evaluating options that includes both rational and emotional factors. Feelings, beliefs, values, and attitudes as well as facts should play key roles in our decisions.

The decision-making process helps you get from what's currently happening to what you want to have happen. Whenever you perceive a gap between where you are and where you want to be — in a job, in a relationship, in the way you spend your time, in any issue — you have the opportunity to exercise choice. The following decision-making procedure is based on John Arnold's seven building blocks to effective decision making.

State Your Purpose

"It doesn't matter how you get there,
when you don't know where you're going."

THE FLYING KARAMAZOV BROTHERS, *jugglers extraordinaire*

All too often, the early part of the decision-making process is neglected. To make the best decision, you need to know clearly what you want. You need to give yourself a target.

Without this sense, you might spend your time thinking about the wrong problem. Formulate the purpose of your decision by putting it in a single, concise statement:

Determine the best way to _____.

Consider an overworked manager who is spending most of her evenings and weekends just trying to keep up with her workload. She realizes that if she continues at this pace, she will burn out. To avoid this condition, she must make a decision.

She could try to determine the best way to get an assistant. That might help out in getting the work done, but it bypasses the larger issue of eliminating nonessential work. She could try to determine the best way to eliminate some work, but that would miss the issue of working more efficiently. She could try to determine the best way to work faster, but that overlooks the issue of delegation.

The statement of purpose — figuring out just exactly what needs to be determined — should be as broad as possible. A narrow, specific purpose at the beginning of the process can miss the essence of the situation and restrict the range of solutions. If the manager tries to determine the best way to get the work done, she's more likely to find a better solution.

Consider some other examples of statement of purpose.

- A teenager has an acne problem. If his purpose is to determine the best way to get rid of pimples, he focuses on the smaller picture. If his purpose is to determine the best way to make himself as attractive as possible, he opens up other possible solutions, such as getting better clothes, getting a good haircut, and losing some weight.

- An executive receives a job offer and must choose between staying at her current job or switching to a new one. A good statement of purpose could be to determine the best way to choose between the two jobs. She could also use this situation as an opportunity to determine the best way to get her ideal job.

- A designer gets a new job on the other side of town and resolves to get a new car. But then she reconsiders the situation. Instead of determining which car to buy, she tries to determine the best way to get transportation. She considers car pools, taxis, leased cars, and bicycles. With this wider purpose, she gives herself a lot more options.

- Two people have an argument. One person considers possible directions of action. He could try to tell the other person off. He could try to force his point.

He could try to determine the best way to come to a mutual understanding.

• A couple is renovating their living room and dining room for Christmas. When they realize there isn't enough time, they adjust their target to find the best way to make the rooms as attractive as possible.

By first thinking about the purpose of your thinking — what you want to end up with after having made your decision — before you decide on alternatives, your decision will be more effective. You can apply this when you need to think about home budgets (determining the best way to improve cash flow), about business marketing (determining the most productive advertising strategy), and about human resources management (determining the best way to improve morale and productivity).

Formulating a broad statement of purpose also helps you distinguish between means and ends. Ends are ultimate objectives, final results, and outcomes. Means are ways to get to objectives. They represent the activities and techniques that help you reach ends. Something can become a means or an end depending on the context of your purpose. For example, a job can be a means towards personal fulfillment and financial security, and it can be an end in the search for a job.

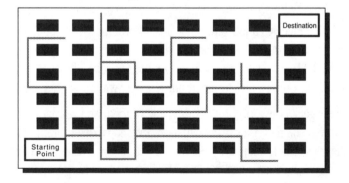

TIP When you are confronted with a decision, ask yourself what the final outcome should be. If you don't have an ultimate end, if you don't know what your purpose is, you can think about means forever.

Establishing Criteria

To clarify your statement of purpose, determine the criteria that will fulfill that purpose. Ask yourself:

- ❏ What do I want to achieve?
- ❏ What do I want to preserve?
- ❏ What do I want to avoid?

Your responses to these questions become a yardstick that will measure how well possible alternatives stack up against each other and ultimately lead you to your best course of action. For example, if you have as a purpose to determine the best way to buy a car, you might arrive at the following list of criteria.

- **ACHIEVE:** A dependable car, with a good repair record and a long warranty. Get a sporty car, one that makes people's heads turn. Get a comfortable car, which provides lots of space for passengers. Get a responsive car, which has plenty of power and good handling.

- **PRESERVE:** Your savings. Get a car with a low price tag and good financing. Your budget. Get a car that has good gas mileage, low insurance premiums, and a good resale value.

- **AVOID:** Costly repairs. Parts should be readily available and reasonably priced. The car should perform reliably.

These criteria form the basis of your decision, so it's important to make them specific. You could set an upper price tag of $15,000. You could specify that the car must be bright red. You might say that the car must have an EPA rating of more than thirty miles to the gallon. You might stipulate that the car be flashy. You could require that the car carry four people comfortably. In this step, you define and develop the standards that describe your statement of purpose.

Examine your criteria to determine whether they contradict each other. Do you want a luxury car for under $3,000? Do you want a car with tremendous performance but very high gas mileage? Do you want a sleek sexy sports car that is also a station wagon?

Once you've settled on an assortment of criteria, set priorities. To each criterion, assign a number from one to ten, with ten representing the most important criterion, and one representing the least important criterion. If, for example, performance is most important to you, it rates ten. If the cost of the car is less important, then it might get a rating of eight. If repair costs seem about half as important as the performance of the car, then it gets a rating of five.

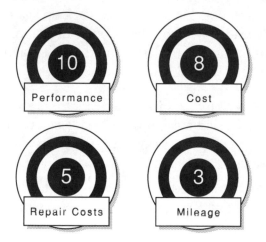

In this rating phase, you determine what's important and what's not so important. Select at least one absolute requirement — a criterion so important that it must be satisfied by any decision — and give it a rating of ten. Rank the other criteria as desirable objectives which would be nice to achieve, but are not critical. In this process of weighing, judging, sorting and sifting through criteria, you may find that the cost of the car becomes more important than the mag tires, or that the warranty becomes less important than the air conditioning.

At the end of the process, you should have a list of factors that specify your objective and ultimately define what you want.

Searching for Alternatives

After you have determined your criteria — what you want to achieve, preserve, and avoid — you need to think about possible courses of action. Find out how you can meet these criteria.

The key to finding alternatives is to let your criteria generate your options. For example, to search for possible cars, ask yourself what car costs less than $15,000, is sporty, yet spacious, has a five-year warranty, and comes in bright red. With this question in mind, you can look in newspapers, magazines, showrooms, advertisements, consumer reports, and build a list of cars that will satisfy at least some of these criteria.

In examining alternatives for a college, you might try to find a school that has a very high academic rating, a great variety of campus activities, many opportunities for research, and a tuition fee of less than $10,000. In searching for a home, you might want to find a three-bedroom fully detached house in a certain area, with a large yard, renovated kitchen and bathroom, under $150,000. In searching for a fulfilling occupation, you may look for a job that's within walking distance, which lets you work with computers, where you don't have to spend any more than forty hours a week, and that pays more than $40,000 a year.

State your alternatives clearly. Write them down. In the long run, well-defined distinctions between alternatives make it easier to make a choice.

TIP Too many times we choose alternatives before we set our criteria. Let your criteria generate your alternatives.

Evaluating and Testing Alternatives

"Swift decisions are not sure."
SOPHOCLES, *5th-century B.C. poet and dramatist*

"He who considers too much will perform little."
JOHANN VON SCHILLER, *18th-century poet and dramatist*

Having listed your alternatives, the next step in the decision-making process is to compare how each alternative stacks up against the others. In some cases, a single alternative may emerge as the clear winner — you may find a car that satisfies all your criteria. But in other cases, each alternative holds advantages and disadvantages. What then?

In situations like these, you need to rank your alternatives. Compare each alternative to each criterion. If an alternative best meets a criterion, give it a rating of ten. It doesn't have to be perfect, just the best that's available. If an alternative is half as good as the best for meeting a criterion, give it a rating of five. Once you do this, you may arrive at a chart similar to the one below.

	Performance	*Repairs*	*Cost*	*Mileage*
Civic	9	6	10	8
Fiero	10	5	7	6
Toyota	9	6	6	10
Escort	6	10	9	7

Next, you multiply the value you assigned a criterion by the rating you assigned an alternative. For example, the criterion of performance received a value of ten, the Civic had a rating of nine, so it scores ninety points. Do this for all criteria, then total the score for each alternative, and see which one is greatest.

	Performance (10)	Repairs (5)	Cost (8)	Mileage (3)	Total
Civic	9x10=90	6x5=30	10x8=80	8x3=24	224
Fiero	10x10=100	5x5=25	7x8=56	6x3=18	199
Toyota	9x10=90	6x5=30	6x8=48	10x3=30	198
Escort	6x10=60	10x5=50	9x8=72	7x3=21	203

In this case, the Civic won as the alternative that best suited the ratings of the criteria. You may want to examine how your decision feels to you. If it doesn't feel right, then maybe your criteria are not ranked as they should be, or maybe you still need to search for more alternatives.

This rating scheme can be applied to feelings as well as specific factual data. You simply incorporate your feelings into a criterion and rate them as you would rate the importance of any other criterion. In this way, you can work towards an alternative that satisfies both your mind and your heart.

Once you have made a selection, it's important to take some time to troubleshoot it. Ask yourself, "What could go wrong?" Then ask yourself, "What could I do to prepare for this going wrong?"

TIP Unless new information and new alternatives appear, carry through the decision that you have made. Avoid the tendency to switch just before the moment of action.

Decisions, Decisions, Decisions...

❏ **Decision-making Warm Ups**
What was one of the best decisions you've ever made? What was good about it? How did you arrive at that decision?
What was one of the worst decisions you have ever made? What was bad about it? How did you arrive at that decision?
What was one of the most difficult decisions you have ever made? What made it difficult?
What is one of the most unusual decisions you have ever made?
What was one of the earliest decisions you made as a child?

❏ **Target Practice**
For each of the following activities, determine a statement of purpose. What could be the end result of these endeavors?

Buying a house
Reading a magazine
Running a mile
Doing all the exercises in this book
Buying lottery tickets
Dieting
Not expressing negative emotions

❏ **Setting Criteria**
Figure out the most important criteria in making a good choice for the following activities. What might you want to achieve, preserve, and avoid?

Selecting a dictionary or atlas
Buying a sofa bed
Selecting a new winter coat
Choosing a novel
Choosing a career
Getting married

Making a Point

Get a piece of paper and rate the importance of the following criteria for buying a house: total cost; mortgage rate; neighborhood; proximity to schools, churches, and shopping; street noise; taxes; spaciousness; arrangement of kitchen; number and size of bathrooms; amount of closet and storage space; space for entertaining; condition of plumbing, wiring, and heating system; age of roof; number of bedrooms; condition of roof, gutters, windows, or foundation; air conditioning; parking; size of yard.

Getting There Is Half The Fun

Determine what activities you may need to perform in order to achieve the following set of goals.

To set up a hockey pool with twenty people in four weeks

To be able to program in PASCAL by the beginning of June

To travel to Kenya by next March

To renovate the living room and dining room by Christmas

Choose and Feel

There is a popular story about Albert Einstein. When confronted with a decision that had two possible alternatives — a yes-or-no decision or a this-or-that decision — Einstein would take out a coin, assign one alternative to heads, the other to tails, and flip the coin. When the coin landed, he would look at the face of the coin and immediately ask himself how he felt. If he felt good, he would go with the first alternative. If he felt bad then he would go with the other alternative.

❏ **Think about the following hypothetical situations:**

If you were going to die tonight, what would you most regret not having told anyone?

If you were offered one million dollars to leave the country and never return, what would you do?

Would you kill a cow if you had nothing else to eat?

If you knew that there would be a nuclear war in five days, what would you do?

Would you choose to know the precise instant of your death?

A special device has been invented that can tell you the correct answer for any single question. Assume that you can ask about anything in the future, the past, or the present. What question would you ask?

If you could live in any person's body for a full day and night, whose would you choose and what would you do?

If you could inhabit an animal's body for a day, what animal would you choose?

❏ **Risk Muscle**

Next year, you must plant one crop over all your land. You have a choice of three kinds of seed. There will be three possible weather conditions, good, okay, and poor. It is impossible to predict which condition will occur, since each is equally likely to happen. The profit and loss for the possible weather conditions are outlined in the following table.

	Good Weather	Okay Weather	Poor Weather
Seed one	24	0	-6
Seed two	12	6	0
Seed three	6	6	6

Which crop would you plant? Why?

❏ **Finding More Alternatives**

The I Ching, also called the book of changes, is a collection of wise sayings that has been read in China for several thousand years. When you have a difficult or complex decision to make, consider using the I Ching, not as a means of divination, but as a way to get your thinking traveling in new directions.

❏ Ranking Routine

Rank the following seven shapes in order of preference. Do this by choosing the shape you like best and the shape you like least and set them aside.

Then, of the remaining five shapes, choose your favorite and least favorite shape, and set them aside. From the remaining three, once again choose the shapes you like best and like least.

❏ The Root of the Problem

Next time you're feeling depressed, take some time to proceed through the formal decision-making process. Ask yourself what the problem might be. Formulate a statement of purpose. Define some criteria. Find out what you need to make yourself feel better.

Default Decision Making

*"If you don't get what you like
then you'll be forced to like what you get."*

GEORGE BERNARD SHAW, *playwright and essayist*

Many decisions — keeping the same job, not exercising your body, eating the same foods, not taking risks — are made by default. These unintentional choices can have as much impact as deliberate decisions. They can cause you to drift through life, carried passively by the ebb and flow of circumstance.

When you consider that many people spend more time planning their Christmas parties than planning their lives, you realize that we can become so focused on daily matters that we forget to look at the big picture.

Having and working towards personal goals is an importance facet of mental fitness. Goals are targets that focus your time, energy, and creativity, and they lay the groundwork for personal growth, filling the basic need of having a direction.

What do you want in life? Do you want lots of money, a big stereo, a beautiful house, fame, glory, a large family, happiness, a long life? Do you crave adventure, health, peace, knowledge? Do you yearn for love, security, excitement, power? Take some time to do the following exercise to reassess your direction in life.

Wish List

Take a piece of paper
and write down your wants.

Write down anything
you want to have,
to see, to do, to be,
or to experience.

Organize these wants into the following categories:

Material Wants
Financial Wants
Career Wants
Recreational Wants
Relationship Wants
Personal Growth Wants

Be daring.
Include secret desires
as well as long-term wants.
Write down big wants,
as well as little wants.

Keep on writing
until you have nothing more
to write down.

Take some time now,
get a piece of paper,
and identify anything and everything
you wish for.

When you have listed all that you want, take some time to examine these wants.

- Set a priority to each want. Which are the most important?

- When did you acquire this want?

- What experiences have contributed to your having this want? If you didn't have these experiences, do you think that you would still have this want?

- Who might want you to have this want?

- How much satisfaction and happiness have you received from having this want?

- How good do you feel from doing things that satisfy this want? How bad do you feel from failing to fulfill this want? How much satisfaction will you ultimately get from having this want?

- What would it be like if you didn't have this want? Would you be a different person? What would you be doing right now?

- Does this want conflict with any of your other wants?

- Do you want to be the kind of person that has this want?

Answering these questions can change the way you view your wants. You may decide that a want is superficial, is unthinkingly adopted from someone else, or is a deeper genuine want. You may want to add to or delete wants from your list. You may want to combine wants into simpler, broader wants, like the wish to be happy no matter what happens. You may find that what you really want is to find out what you really want. Maybe you want to be able to control what you want.

Feel free to spend lots of time thinking about your desires. After all, these urges steer your course in life.

Many of our wants can be traced back ultimately to the want for approval from others. The wish to be attractive, to be powerful, successful, and wise can be powered by the wish to be admired. What do you gain from others' approval? Perhaps it makes you feel justified in approving yourself. However, wanting approval from others leads to giving up your freedom to the will of others.

Transform each major want into an objective. Determine the best way to fulfill this want. What do you want to achieve, preserve, and avoid to satisfy this want? Use these criteria to generate activities. Rate which activity would best fulfill this want.

Finally, put these activities into specific goals. Do this by making an activity specific, measurable, and tied to a date. For example, if you want to be rich, then consider the long-term goal of having $200,000 in your bank account in ten years' time, and the short-term goal of establishing a part-time company that will bring in $1,000 a month by next April. If you wish to be physically fit, then set the goal of working out three times a week, for the next three months.

Realistic goal setting is crucial if you are to get from where you are to where you want to be. If you find that a want will not translate easily into a goal, then you should examine how much you really want it. Form short-term and long-term goals. Make an objective challenging, but not so difficult that you lose motivation. Make the objective realistic, but not so easy that it becomes trivial.

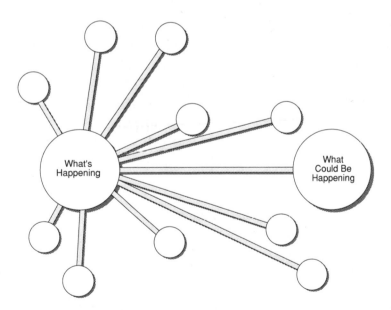

TIP Happiness is not a goal to be achieved, but a result of achieving goals.

Decision Making Style Tips

Good decision making can be the key to your mental fitness. Take time to develop good form in your choices. It will increase the coherency of your plans and the effectiveness of your actions.

TIP ONE Become more aware of your decisions. Recognize which of your choices are made from habit and which are made from conscious analysis.

TIP TWO When confronted with a decision, state the purpose of your thinking. Use the KISS Rule — Keep It Short and Simple — to zero in on your target. Remember that a poorly defined problem can have countless solutions.

TIP THREE Establish the criteria on which you will make your decision. Break down your choice into several components which you can use as a basis of comparison rather than a few broad brush strokes. Distinguish between alternatives.

TIP FOUR When you consider an important decision, consult your thoughts and your feelings. Try to arrive at an alternative that feels right. Always remember to keep sight of your ultimate goals. Once you have make your decisions, accept them.

"The purpose of life
is a life of purpose."
ROBERT BURNS, *18th-century poet*

11

IMPROVISING

The Creative Act

On the Spot

Take a couple of moments
to relax and unwind.

Let your attention wander freely,
allowing it to glide from one thing to another.

When your mind is poised
and you feel ready,

do something creative.

Go ahead and do it.

Start now.

The Creative Process

*"Creative thinking may simply mean
the realization that there's no particular virtue
in doing things the way
they have always been done."*

ROGER VON OECH, *creativity consultant*

Creativity is at the heart of intelligence. Yet trying to define it is like trying to hold down and capture a drop of quicksilver with your fingertip; the moment you think you've got it, it breaks apart into dozens of drops that roll off into different directions.

Does an idea have to be entirely original to be creative? Must something be artistic or aesthetically pleasing before it's creative? Can creativity result from detailed methodical work? Is creativity always accompanied by a feeling of inspiration, a spark of insight, or a sense that the idea came out of nowhere?

Some psychologists think that creativity is nothing more than innovative problem solving. From this point of view, there is nothing extraordinary about creative thinking. It happens as a consequence of methodical thought. Other psychologists think that creativity is a kind of wild, uncontrollable process that taps the unconscious mind and involves great leaps of insight. From this point of view, creativity is mysterious and unpredictable. There is some truth to both views.

Creativity involves having an idea yourself rather than borrowing it from someone else. You witness the birth of the idea in your mind. This can happen through persistent attempts to solve a problem, say by applying a series of well-defined steps to a project, or it can happen as if by magic, when an inspiration appears suddenly and unexpectedly, seemingly without any volition on your part.

In the opening exercise, you could have done any number of things — from standing on your head, to making jungle sounds, singing a song, tapping your fingers in rhythmic patterns, or thinking about quantum physics. The point of the exercise was not to make you have a creative breakthrough, but to get you thinking about what creativity means to you.

A creative act can solve a problem, serve a specific purpose, or fill a functional need. Alternatively, a creative act may not serve a practical purpose, but fill an emotional or aesthetic need. Whether creativity is applied to designing a bridge, inventing a recipe, painting the Sistine Chapel ceiling, or writing an autobiography, skill and diligence are required to to produce an innovative product. In the words of psychologist Abraham Maslow, "A first-rate soup is more creative than a second-rate painting."

The creative process can be divided into two main phases: exploration and application. In the exploration phase, you generate and manipulate new ideas. You bring unrelated material together, make fresh connections, and look for unusual patterns. You imagine, fool around, break the rules, and let ideas simmer in the back of your mind.

In the application phase, you judge and implement your ideas. You determine how applicable the idea is, whether it meets your criteria. And you launch the idea into action, taking it from "what if" to "what is." The two phases are complementary: during the generation phase, you widen your thinking; during the application phase, you narrow your thinking.

In creative thinking, as in high jumping, timing is everything. If you try to be practical, cold, and logical in the exploration phase, you'll focus on limitations rather than on possibilities. Similarly, if you are free and associative during the application phase, you may not get your idea into action, or you may not see the pitfalls of your idea until it's too late. Know when you need to focus, and when you need to loosen up.

"Creativity is ten percent inspiration,
and ninety percent perspiration."

THOMAS EDISON, *inventor*

Look Beyond the First Right Answer

*"The best way to get a good idea
is to get a lot of ideas."*

LINUS PAULING, *Nobel physicist*

EXERCISE What is this?

Much of our thinking is geared to towards finding a single right answer. With this "one correct solution" approach, instilled through exams, multiple choice tests, and short answer quizzes, we define boxes of yes and no, correct and incorrect, black and white.

A consequence of this style of thinking is that when we are looking for new ideas, we stop at the first right answer and go no further. The first solution, even though it may not be the best solution, blocks the urge to keep on looking. We lose the opportunity to find an even better solution.

For example, you could think of the above drawing as two circles and stop there. Or, you could take it further and think of it as a fried egg, the top of a toilet paper roll, a sombrero from below (or above), the eyeball of an albino, a faucet washer, the orbits of Mercury and Venus, the bottom of a light bulb, or a billiard ball for the CIA (the numbers are classified).

So when you're in a position to do something creative, keep looking. It may be that nineteen out of twenty ideas can be shelved permanently, but remember that Thomas Edison found over two thousand ways not to make a light bulb. If you don't continue the search, you may find a better solution.

TIP Don't stop at the first idea. Keep looking.

Making Connections

"The ability to relate and to connect
sometimes in odd and yet striking fashion,
lies at the heart
of any creative use of the mind,
no matter in what field or discipline."

GEORGE J. SEIDEL, *author*

EXERCISE   What do you get when you cross a light bulb with a pen?

Many creative advancements have happened because someone combined two seemingly different ideas and turned them into something new. Johann Gutenberg combined the idea of a grape press and a coin punch to make the moveable type printing press. William Harvey made a connection between a water pump and the human heart and developed the modern theory of circulation. The Wright brothers made a connection between a bicycle, a bird's wing, and an engine, and were the first to enjoy powered flight.

In the above exercise, you may have made several connections. You may have thought of a pen with its own built-in light source for writing in dark places. Or you may have thought of a special pen which can write on large bulbs and cause pretty patterns to be projected. Or you may have thought about a pen that screws into a holder and is refilled. What other connections can you think of?

It's interesting to watch the connection take place in your mind. At one moment, you're searching, and the next moment, you've found an idea. What happens in between? You experience an "aha." The "aha" you feel when you make a connection is similar to the "haha" you experience when you find something funny. Both are spontaneous.

TIP   Look for things to connect with your idea.

Trigger Words

Imagine that your mind is like a huge library, and that your memories, experiences, and ideas are the classics, romances, pulp fiction, science texts, plays, and comedies that stack the shelves. Many of these works are hidden away in dimly lit corridors. Their ideas are available, but not easily accessible.

There are several good ways to improve access to your mental library of experiences. One is to use trigger words — concepts that stimulate fresh associations. Browse through a dictionary or thesaurus and let your mind find associations between the words and the idea that you're working on. You can look until you find a word that sparks a new association or you can close your eyes, open the book at random, place your finger somewhere on that page, and force a connection between the word your finger landed on and the idea you're working on.

For example, say you're developing a new board game, and you land on the following word.

ENZYME

What could this mean to you? An enzyme is an organic substance that causes a chemical transformation. The word could remind you of cells and microbes and make you think about a game in which some players are microbes and try to infect a cell, and other players try to protect the cell. The word may make you think of one large cell or of an entire organism. Since enzymes assist in chemical reactions, but are not used up in the reaction, you may think of a game where a player's position doesn't change, only the environment is altered. Thinking in different directions, the word *enzyme* sounds like the word *ensign*, and may make you think of military battles.

How could you connect the following words to the game concept: *hallway, prince, light bulb, cartwheel, bunion?*

Metaphor and Simile

"The greatest thing by far
is to be a master of metaphor."

ARISTOTLE, *4th-century B.C. philosopher*

EXERCISE Think about the following similes and give reasons why they could apply.

• Creativity is like baking a cake.

• Creativity is like falling down in the mud.

• Creativity is like making love.

• Creativity is like fixing a leaky faucet.

• Creativity is like sharpening an ax.

Another good way to make connections in your mind is to use similes and metaphors. Both similes and metaphors make comparisons and highlight similarities between two things. Similes are easier to recognize because they contain the words *like* or *as*. Our language is filled with metaphors. Chairs have arms and legs, clocks have faces, puzzles have keys, thoughts are clear, and computer programs can be user friendly.

Metaphors and similes can lead your thinking in new directions. If you think of creativity as being like baking a cake, you might think that ideas must be mixed together in the correct proportions, need time to rise, and have to be baked at the right temperature.

Make a mental note each time you discover a simile or a metaphor. Look for them in conversations, television shows, movies, novels, and articles. Ask yourself whether they makes sense. Do they appeal to your senses, to your feelings, or to your intellect? Next time you need to find a new approach, try putting your idea into a simile or a metaphor.

Divide and Unite

Another good way to generate ideas is through an approach called *attribute analysis*. In this technique, you list the features, characteristics, and parameters of the idea that you are thinking about. Then you alter one or more of these characteristics to create a new idea.

For example, imagine that you're working on a better cup design. As a list of the cup's attributes, you might decide that cups are round, they have handles, they are made from something solid, they have a steady base, and they don't change a liquid's taste.

Now you're ready to think in new directions. Zero in on one attribute and alter it in some way. For example, give a cup two handles instead of one so that it can be grabbed from either side. If you put the handle on the inside you would never drink anything that's too hot for your mouth, since your hand touches it first. You could make the handle as large as the whole cup, for people who drink wearing hockey glovers.

Change other attributes and you make the cup serve specific purposes. Give it a pointed bottom and the cup won't fall over in the sand at the beach. Make the bottom of the cup soft and sticky and it won't tip over on rough ocean voyages. If you want to make a cup for people who drink too much coffee, strategically place a hole halfway up the side. That way it will never hold a full cup. For people who like their drinks sweet, chemically coat the inside of the cup with sugar. Coat the cup's rim with four different flavors — sweet, salty, sour, and bitter — and you can dial a taste by turning the cup around.

TIP Try altering one aspect of an idea at a time.

Change Contexts

EXERCISE Think of twenty possible uses for a paper clip.

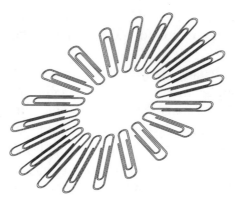

Things get their meaning from the context in which they are placed. For example, in an office setting, a paper clip is used to hold together letters, notes, and memos. But in a laboratory, the same curl of metal can act like a wire and connect two loose cables. On a pair of broken glasses, it acts as fastener. For a thief, it picks a lock. With a piece of tissue paper, it can clean out the inside of a Swiss army knife.

By changing context, you explore the possibilities of your resources. When you're traveling, your towel becomes a pillow, a sun shade, a food drier, a sand mask, a shirt, a camera bag liner, a laundry basket, and a backscratcher.

> *"Make it a practice to keep on the lookout*
> *for novel and interesting ideas*
> *that others have used successfully.*
> *Your idea has to be original*
> *only in its adaptation to the problem*
> *you are working on."*
>
> THOMAS EDISON, *inventor*

Be Open

*"The creative process cannot be summoned at will
or even cajoled by sacrificial offering.
Indeed, it seems to occur most readily
when the mind is relaxed
and the imagination roaming freely."*

MORRIS KLINE, *mathematician*

In Greek mythology, it was thought that the nine daughters of Zeus, called the Muses, breathed creative ideas into favored artists and musicians. In fact, the word *inspire* comes from a Latin word meaning breathe in.

Beneath your consciousness, your mind is engaged in an elaborate process of computing, sorting through, arranging, and rearranging information. When a connection is made, it travels through your system like an electrical spark.

When you're working hard on a project and you want a fresh jolt of mental energy, you should take a break. Do something completely different. Go for a walk, listen to some music, read a magazine, browse in a bookstore. Set aside half an hour and return to the project later. This gives your subconscious mind time to work on the idea, as well as your conscious mind the repose to continue to develop the idea.

Sometimes, when our subconsious mind has free reign, such as when we dream at night, its connections are more interesting than those our conscious mind makes. Because deliberate thinking can get caught in repetitive patterns, it loses its freshness and spontaneity. To explore one way to engage your subconscious mind in creative thinking, try the following exercise.

Creative Non-Thinking

Sit down for a few minutes,
and find a comfortable posture.

Make your mind calm and quiet,
and your attention receptive.

For a few moments,
think creatively
about how to make yourself
more attractive.

For five minutes,
hold that question
in your mind,
but don't think of any possible solutions.

If you start thinking about answers,
push those thoughts from your mind.
You may want to explore further in a direction
suggested by a thought, but don't.
Maintain the question,
but don't look for answers in the usual way.
Be ruthless.

Aim for a pure, mental state of questioning.

You may find this difficult at first.
But by rejecting all initial answers,
and thinking in a non-habitual way,
you force yourself to find something genuinely new.

You engage regions of your mind you may not use very often.

Remember that Nature abhors a vacuum.

Improvisation Exercises

❑ **Warm-up Questions**

What was the last creative idea that you had?

When was the last time you put a creative idea into action?

What are the five most creative things that you have ever done? Think back to your childhood.

What was the last creative risk that you took? What happened? What was the worst that could have happened? What was the best that could have happened?

❑ **Mental Flexors**

You are given a candle, a book of matches, and a box of tacks, and must attach the candle to a wooden door so that the candle will burn properly and will give off enough light to read. Pay attention to where your ideas come from as you think.

List as many ways as you can think of to measure the height of a building using a barometer.

You are applying for a job as creative director for an advertising company. On the morning of your interview, you plan to put a message somewhere the president will see on his drive to work. What creative message would you produce? How would you present it so the president is sure to see it?

You know that you will be given a test of getting a ping pong ball out of a long narrow cylinder that is bolted to the floor. The usual solution is to pour water into the tube and let the ball rise up. But since you know in advance, you prepare by taking some extra materials. What would you take, and how would you use it to get the ball out?

❑ **Practical Creativity**

Choose one routine that you perform regularly — it may be cooking a meal, giving a presentation, writing a monthly report — and develop one way to make it more creative.

Metaphorical Muscles

Fill in the following blanks and complete the similes and metaphors.

Water is to a ship as ... is to a business.
A flower is to joy as ... is to anger.
A faucet is to ... as ... is to freedom.

My house is ...   My job is ...
My spouse is ...   Anxiety is ...
Truth is ...   Love is ...
Power is ...   Ideals are ...
Thinking is ...   Success is ...
Happiness is ...   Life is ...

Synthesia Muscles

An interesting way to enter into a playfully creative frame of mind is to imagine your modes of perceptions crossing over each other, so that you would taste sounds, hear colors, and smell sensations. Try imagining these perceptions.

What does the word *participate* smell like?
What does the number seven feel like?
What does the color blue taste like?
What does the idea of freedom look like?
What shape is Wednesday?
What does joy taste like?

How would you:

Develop a group sport that involves two balls.
Redesign the human body.
Redesign the human face.
Design a house that has no straight walls.

Everyday Creativity

Every day, put your body into a position you have never put it in before.
Every day, make up a new word, and invent the meaning of the word.
Every day, think up a new thought.
Make up a mental exercise every day.
Try brushing your teeth in a different way each day for a month.

❏ Spontaneous Illustration

Take a pen and doodle. Without having an idea of what you're going to create, press the tip of your pen to the paper, and allow it to draw. Don't be concerned with what the outcome will be. Just let it happen. Draw at least fifteen pieces. What does your pen draw?

❏ Dream Programming

Dreams can often be a source of creativity and imagination. When you want to think about something, loosely hold a picture of it in your mind as you drift off to sleep. Instead of seeing the picture directly in front of you, imagine that the image is off to the sides, waiting to appear in front of you when you drift off. When you wake up in the morning, recall what you have been thinking about, and write it down. Splash your thoughts onto a piece of paper at your bedside.

❏ Two-Minute Creativity

Very often, our creative juices circulate best when we're put on the spot and forced to be inventive. Nothing loosens you up better than having to do a lot in a short period of time. Perform each of these exercises in two minutes.

Do as many things with this book as you can in two minutes.

Put your right hand in as many different positions as you can in two minutes.

Say the word "tonight" in as many different intonations as you can in two minutes.

Here are three shapes. Draw as many designs as you can using only these shapes for two minutes.

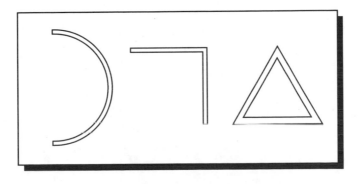

Creativity Style Tips

Creativity is that remarkable quality which makes us innovative and thoughtful. It is the ultimate aim of the artist in our mental athlete.

TIP ONE Look beyond the first right answer. Make it a habit to search just a little longer and to dig just a little deeper.

TIP TWO Make connections. Use metaphors, similes, and analogies to generate new ideas. Look for material outside of your usual sphere of interest.

TIP THREE Take creative risks. If you're going to get original ideas — original for you — you need to do something different than you usually do. This means venturing into the unknown and attempting something new. Don't be afraid to put on your mental safari suit.

TIP FOUR New insights, inspirations, and creative leaps of imagination can appear any time, any place — in the shower, over lunch, when you awake in the morning. And very often, they can disappear just as quickly as they appear. Be prepared to record them as they arrive: jot down or draw a picture of your inspirations.

"If you do not expect the unexpected you will not find it, for it is not to be reached by search or trail."
HERACLITUS, *5th-century B.C. philosopher*

12

PEAK PERFORMANCE

Learning about Learning

Mental Patterns

With your left hand, tap out a rhythm
between your thumb and fingertips.

Tap once to your index finger,
twice to your middle finger,
once to your ring finger,
twice to your pinky.

Then repeat it backwards,
once to your ring finger,
twice to your middle finger,
once to your index finger,
and twice to your middle finger,
and so on.

Repeat this pattern continuously several times,
so it becomes automatic.

Then with your right hand,
try another pattern:
Tap once to your index finger,
twice to your middle finger,
three times to your ring finger,
twice to your pinky,
once to your ring finger,
twice to your middle finger.
and so on.

Then perform both patterns simultaneously.

Pay attention to what it feels like
to learn how to do this.

Getting the Knack

"I hear, I forget.
I see, I remember.
I do, I understand."

CHINESE PROVERB

How would you learn how to swim? First, before you do anything else, you have to put yourself into the water. You can jump in all at once or you can slowly inch yourself in — first testing the water with your big toe, then getting the bottom of your foot wet, then your ankle, calf, your other foot, your knees, thighs, stomach, chest, shoulders, neck, and finally your head.

Once in the water, what's next? You try to swim. But how? To start, you need a routine, a point of departure, a technique. You can start by watching someone who already knows how to swim and then mimic that person. You can lunge into the water and paddle. You soon realize that you need to kick your legs. You figure out when to breathe by discovering when not to breathe. You push yourself forward by pulling the water beneath and past your body. Sooner or later, amidst all the splashing, *you get the knack.* Your arms move with your legs. Your breathing becomes properly paced. You slice through the water in one fluid motion.

To swim better, you practice. You watch more people and observe what they do to swim longer and faster. You ask for tips and suggestions from others who have more experience. You experiment, try out new strokes, and continue to practice. With patience, some effort, and still more practice, you learn.

The basic approach to learning new things — whether it's figuring out how to cook a soufflé, how to dance a polka, how to invert a spreadsheet, how to speak in front of an audience, how to play the saxophone, how to write a novel, or how to swim — is the same: trial and error.

What did you observe when you attempted the opening exercise? Did you find that it was difficult at first, and then progressively easier? Did you witness yourself getting the knack? Did you find that some of your emotions — perhaps frustration or boredom — interfered with the learning process? Did you experience yourself learning?

Learning is a process of creating new habits. Each time you grasp a new idea or skill, you establish a pattern of connections between your nerve cells. As you learned the finger tapping exercise, you established a complex pathway of neural circuitry.

When we learn, we are literally programming our brain. Complex skills — such as singing, writing, dancing — are built on successive layers of habits. To play the piano, you develop habits of holding your hands, of striking the keys, of playing scales, and of playing songs. When you become proficient at one level your attention becomes free to work on a higher level. The virtuoso pianist is free to concentrate on the emotional expression of his playing rather than on where his fingers are.

*"If you want to do something,
make a habit of it;
if you want not to do something,
refrain from doing it."*
EPICTETUS, *1st-century philosopher*

Creating the Need

QUESTION What motivates you to learn?

- ❑ Curiosity
- ❑ Sense of accomplishment
- ❑ Fear of failure
- ❑ Money
- ❑ Dissatisfaction
- ❑ Fun
- ❑ Learning for learning's sake

Whether we like it or not, we are stimulus-response creatures. Our behavior is shaped by basic motives to be rewarded and to avoid punishment. These motives can be external — financial incentives, the knowledge that if we don't produce, we'll be fired. The motives can also be internal — an insatiable curiosity, a fear of being left behind, an urge to have fun, and a wish to grow.

Giving yourself a reason to learn tends to make you progress faster. A good way to encourage learning is to put yourself in a situation where you must progress. Do something small that forces you to learn something big. It could be signing up for a course that has essays, tests, and final exams. It could be investing money in a stock portfolio. It could be offering to give a talk or teach a course. Or it could be buying a book and feeling that you must read it. Necessity is the mother of learning.

My friend John tried everything to stop smoking. Finally he decided to give his lawyer several registered letters that outlined all the nasty, embarrassing things that John had ever done. If he couldn't stop smoking by a certain date, the lawyer was instructed to send the letters to John's friends and coworkers. The fear that the dirty laundry would get out was enough to make him quit smoking.

TIP Put yourself in a position where you need to perform.

Mental Readiness

"Who learns by finding out,
has seven fold the skill of him
who learned by being told."

ARTHUR GUITERMAN, *poet*

What do childen do when they play with building blocks? They put small blocks on top of large blocks. They put large blocks on top of small blocks. They stack them together by color, by size, by shape. In this playful process, they learn how the pieces fit together and they improve their building skills.

One reason children learn so well is that they haven't developed preconceptions of how they are supposed to learn. They also have not developed the notion that play and work are mutually exclusive activities. Play is an important part of the learning experience. When we have fun learning, we learn better. We adopt an experimental attitude by trying something and seeing what happens. Fun is a powerful motivator. It is infectious and spreads like wildfire. It makes people more productive and helps them feel more fulfilled.

We learn when we are in the mood for learning. To learn better, take some time to prepare your mind. Whet your appetite. Focus on the interesting aspects of a subject. Spark your imagination with questions. Consider some of the following ways to help ready your mind.

- When you need to study a subject from a textbook, take a couple of minutes to prepare your mental muscles and to get your mental juices flowing. On a piece of paper, write out, as quickly as you can, what you already know about the subject. Use the clustering technique described in Station Five to splash out your thoughts. Jot down what you studied the last time. You will discover blanks and holes in your understanding, which you can fill in by studying. Take two minutes to do this.

- When studying from a textbook, give yourself specific objectives of how much you want to learn. Quickly skim the material, get an overview of the major concepts, and decide how many pages you want to study. When you read through the material in depth, find a good pace. Focus on the concepts and meanings, rather than on getting through the pages. After you have finished, review the material. Make your study time a complete cycle of action.

- Discover what you don't know about a subject. A good way of doing this is to visualize the subject in detail and then ask yourself questions about what you see and how things would happen. For example, say that you're studying the way the digestive system works. You can visualize how a hot dog gets broken down in the mouth, how it travels down the esophagus, and what happens to it in the stomach and intestines. As you visualize this, ask yourself what substance breaks the food down. Ask yourself how long the process takes. Ask yourself how nutrients are transferred from the food to blood cells. What would the process look like if you could witness it taking place?

- When we learn we often try to confirm and substantiate our position. We look for material that supports our point of view. We might search for material that expresses other points of view. In this way, we add depth to our understanding. Learning is like looking through a microscope and examining what you see. The purpose is not to convince someone else of what is there. The purpose is to search and to get to the truth of the matter.

- If you become confused or baffled when studying a subject, take some time to establish anchor points. Go back to a place where you felt comfortable with the material, and figure out what specifically is confusing you. Deal with one area of confusion at a time. Make sure you don't skip over an idea if you don't understand it.

- Be ready to learn. You never know when the universe has a lesson for you.

"You can learn a lot just by watchin'."
YOGI BERRA, *baseball player and manager*

Mental Blocks

EXERCISE What keeps you from learning?
- ☐ Laziness
- ☐ Lack of money
- ☐ Fear
- ☐ Habit
- ☐ Lack of direction
- ☐ Nothing left to learn

There are several types of mental blocks that keep us from learning and from moving forward in our thinking. One of the largest and most formidable blocks is the habit of laziness. In one area of life or another, most of us can be idle, passive, or inert. When we become rooted in laziness and unwilling to make an exertion, we don't move on.

Learning new things can involve a confrontation with the unfamiliar and the unknown. When this happens, we can rub up against another block: fear. Many fears are based on imagined perceptions rather than on reality. We can have the fear of appearing stupid in front of other people. This makes us never take a risk. We can have the fear of failure, where we never strive to do anything new. Or we can have the fear of change, and become too comfortable in our rut. The effect of fear is to make us cautious, to protect ourselves from any perceived danger.

If fear stands between you and something you want to learn, the only route is through it. As you approach your goal, your fear will diminish and your confidence will increase. When you analyze what you have gone through, and realize that you are capable, you will feel safe.

In just about anything we do, there's a chance that something will go wrong. Because there are always factors that are beyond our control, and because our knowledge of things is inevitably incomplete, there is a possibility that we will make a mistake. In our culture, we maintain a strong association between making mistakes and being a failure. Fear of making a mistake becomes a block.

Mistakes are opportunities for learning. Errors are essential to the learning process. They tell us when it's time to examine things more closely and to try another approach. When you catch yourself making a mistake, think of it as a transition between not knowing and knowing. You can't make progress without making mistakes. Mistakes can also provide impetus to carry out a change you may have wanted to make for some time, but have lacked the motivation to carry through.

When you've made a big mistake, think of the words of Walt Disney: "You may not realize it when it happens, but a kick in the teeth may be the best thing in the world for you."

Mathematician George Moore once said that to get better at mathematics, you have to practice doing the problems you keep getting wrong. There's no point in practicing the problems you already can solve. Keep track of mistakes you make, and use them as a guide to improving.

Another mental block is the idea that you already know as much as there is to know. Either subtly or not so subtly, we hold on to favorite ideas, beliefs, and points of view. We make our positions unwaveringly firm and defend them with all our energy. With this attitude, no real learning can happen.

Because learning involves change — in viewpoint, in understanding, in attitude, or in approach — we must possess a willingness to refine our positions. We think that we cannot be wrong, so we fail to consider other points of view. If we think that we are the best we can be, we leave no room for improvement. When we think we already know everything about something, we stop the learning process.

TIP Use mental blocks as incentives to progress.

215

Mental Fitness

❑ **Warm-up Questions**

How did you learn to drive a car?
How did you learn to write?
How did you learn to ride a bicycle?
How did you learn to be more aware?
How did you learn to consider other people's opinions?
How did you learn to relax?
How did you learn to get angry when you needed to?
How did you learn to be curious?

❑ **Expanding Mind**

Here are some good questions to get your mind moving.

Why is the sky blue?
Why do whitecaps form on the ocean and other bodies of water, and why are they white?
Why does a piece of chalk produce a terrible squeal if you hold it in a certain way?
Why do Frisbees fly? Do they have to be spinning to work?
Why do fish travel in schools, and birds fly in V patterns?
Why does the moon seem larger at the horizon?
How do microwave ovens cook meat?
Why do stars twinkle?

❑ **The Curious Mind**

It is as important to be able to formulate questions as it is to be able to think about and answer questions. Think of three questions about each of the following subjects.

Astronomy	Psychology
Physiology	Chemistry
Economics	World Peace
Politics	Religion
Biology	Life

Professional Adapting

When you are put into a novel situation and need to perform, ask yourself how a professional would do this. If you are fixing a leaky faucet, ask yourself, "How would a plumber approach this?" If you are being interviewed on radio, ask yourself, "How would a celebrity act?"

Pondering

The mystery of life is not a puzzle to be solved, but a reality to be experienced. To taste the mysteries of life, exercise your ability to ponder. Choose a question or a thought and imagine pasting it on your forehead. Everywhere you look, the question will still be in front of you. Decide how long you want to hold it in your mind.

Going in Circles

Get a pen and several pieces of paper. Try to draw a perfect freehand circle on the paper. Observe your technique for learning to do this. Try holding the pen in a different way. Vary the speed and direction in which you draw the circle. Try holding the paper in a different way. Continue until you get the knack.

What can you learn from the following story?

A student visited a Zen master to ask him about Zen. As is the custom, the master served tea. He poured the student's cup to the brim, and then kept on pouring. The student looked on with astonishment as the tea overflowed and finally he said, "The cup is full. No more will go in." The master stopped pouring, looked at the student, smiled, and said, "Like this cup, you are full of your own ideas, opinions, and speculations. How can I teach you until you empty your cup?"

Vital Learning

What five things that you have learned in your life would you want to pass down to your grandchildren?

List five questions that everyone should answer if they want to consider themselves a wise human being.

217

Don't-know-mind

Look around the room you are now in,
and let your attention wander
from one object another.

After a few moments,
when your eyes settle on an object,
ask yourself,
"What is this?"

Then respond by saying,
"Don't know."

For example,
if your eyes come to rest
on a light switch,
ask yourself,
"What is this?"

Instead of saying,
"This is a light switch,"
say,
"Don't know."

Look at the object again and repeat the question.

Don't-know-mind is an open mind.
It encourages you to see beyond
the labels that filter your perception.

You see and become aware of what you don't know.
Your perceptions become fresher.

Seeing Learning

Take a few moments to relax,
loosen up,
and collect your attention.

Visualize a large white candle
with a bright steady flame
shining several feet in front of you.

Imagine that the flame
is intensely alive and sentient.

Imagine that it possesses
an awareness
or a quality of "knowingness."

Concentrate on the light
of this awareness.

As you gaze at the flame,
imagine that the light
is shining through you,
soothing you,
comforting you,
becoming warmly aware
of all of you.

Imagine that this awareness
knows all that there is to know.

What would it feel like
to be in the presence
of such an omnipotent light?

Mental Proteins
and Mental Carbohydrates

"If the brain sows not corn,
it plants thistles."

GEORGE HERBERT, *17th-century poet*

Just as your body needs food to function, your mind needs perceptions, impressions, feelings, thoughts, and ideas to keep strong. Each day, your mind is subjected to thousands of stimuli. Television, radio, newspapers, magazines, books, records, movies, plays, advertisements, conversations, junk mail, matchbook covers, and countless other stimuli bombard your senses and affect your mind.

Life is like a spectacular restaurant, in which the waiters are constantly bringing you wonderfully exotic foods. Imagine what you miss if you refuse the unusual and opt for the same dish you had yesterday, and the day before, and the day before that.

A balanced mental diet should provide nourishment from the basic food groups. Every mind needs mental proteins, food that builds the structure of your mind. Mental carbohydrates provide quick energy, but don't have any lasting nutrition. Mental fats act as fuel, and cushion and insulate our inner mind.

Sometimes we consume too much of one kind of food and not enough of the others. Does your mind eat too much junk food? Does it spend a lot of time chewing gum? Do you overindulge in mental fat, to the point where you're beginning to feel your mind get heavy and slow down? Are the extra proteins adding too many mental calories that you can't burn off?

Encourage your mind to sample different foods. Instead of spitting out foods that may be new, taste and savor the new and exotic flavors.

TIP Keep a balanced mental diet.

The Journey

"Compared to what we ought to be,
we are only half awake.
We are making use of only a small part
of our physical and mental resources.
The human individual possesses powers
of various sorts which he habitually fails to use."

WILLIAM JAMES, *19th-century psychologist*

Learning is like traveling. You attempt to journey from one place to another, from not knowing to knowing, from incapable to capable. From maps, stories, and photographs, you get a sense of direction and an idea of where you want to go. As you venture across the land, you discover that some of the maps are hopelessly inaccurate and some are remarkably true.

With more time on the road, your sense of direction improves. You know when you're on the right track and when you're not. You learn to recognize shortcuts as well as dangerous places. You learn to construct better maps. You begin to understand new languages and to use foreign currencies. You learn how to travel fast and light when you're in a hurry, and when to travel in style when you take the scenic route.

There is no final end to learning. The more we see, feel, and touch, the more we realize there is to experience. The more we travel, the more we realize just how vast our world is. The more we know, the more we realize what we don't know.

There is a beauty in this realization. When we understand that learning is endless, we become inspired to open our senses and sharpen our mind. We continue with renewed energy to explore the endless landscape of learning.

TIP The destination is the journey.

Learning Style Tips

As we settle into our niche in life, we become increasingly set in our ways. The good aspect of this is that we build routines and develop habits that free us from having to think about what we've done before. The bad part is that we can become entrenched in patterns of thinking, feeling, and doing that can keep us from learning further.

TIP ONE We learn when we want to learn. Put yourself in a situation where you have to perform. Explain something to someone. Make a commitment. Do something that forces you to learn.

TIP TWO We also learn when we have fun. Look for the exciting and the interesting aspects of what you want to learn. Fool around. Experiment.

TIP THREE Encourage yourself to adapt. When you encounter a new situation, be flexible. Keep your mind open and ready to learn.

TIP FOUR Although there are milestones in the learning process, remember that there is no end to learning. The reward in learning is the ability to learn more. The end is the means and the means is the end.

*"I am putting myself
to the fullest possible use,
which is all, I think,
that any conscious entity
can ever hope to do."*

HAL 9000 COMPUTER, *from the movie* 2001: A Space Odyssey

KEEPING FIT

Developing an Exercise Plan

Rate Your Level of Mental Fitness

How would you rate your mental strength?

❑ 1 — What was the question again?
❑ 2 — My attention span is shorter than a television commercial.
❑ 3 — I can concentrate when I need to.
❑ 4 — I'm able to bench press my own weight in thoughts.
❑ 5 — In the league of Newton, Leonardo, and Einstein.

How would you rate your mental flexibility?

❑ 1 — Things are either black or white.
❑ 2 — Sometimes, things can be in shades of gray.
❑ 3 — I can often see many sides of an issue.
❑ 4 — My imagination is supple.
❑ 5 — Mental contortionist extraordinaire.

How would you rate your mental endurance?

❑ 1 — Isn't this questionnaire over yet?
❑ 2 — I can concentrate, unless something better comes along.
❑ 3 — When I have to, I can put out the effort and go the full mile.
❑ 4 — "Persistence" is my middle name.
❑ 5 — Regularly run a mental marathon and don't even get winded.

How would you rate your mental coordination?

❑ 1 — Think I clearly sometimes not really, sometimes, got it?
❑ 2 — When under pressure, I stumble over my thoughts.
❑ 3 — I usually think things through clearly.
❑ 4 — My thinking is like beautiful music.
❑ 5 — Overall gold medal winner in Mental Olympics.

Developing an Exercise Routine

"In the dim background of our mind
we know what we ought to be doing,
but somehow we cannot start.
Every moment we expect the spell to break,
but it continues pulse after pulse and we float with it."

WILLIAM JAMES, *19th-century psychologist*

There's a curious property about things that we know are supposed to be good for us. Even though we may want to exercise, and even though we always feel better *after* we exercise, for one reason or another, we don't get to it. Either we feel that we don't have the time, or we're not in the mood, or we just forget about it.

Every skill takes time to master. To find this time, you need to set priorities. If something is important enough, you'll find the time and you'll know what else to eliminate. Fifteen minutes of mental exercise a day is a small investment when you consider the enormous dividends that it can pay. A quarter of an hour is less than two percent of a full sixteen-hour day.

Remember that you can exercise just about anywhere, and just about any time. At a bus stop, in an office waiting room, in a bank line-up, walking to work. If you develop the habit of using your time to think actively and purposefully — thinking through the options of a decision, holding something in your attention, or just relaxing — your mind becomes more limber.

As an experiment, for two weeks set aside fifteen minutes a day and work out your mind. It may be just before breakfast, or at lunchtime, or just after dinner. You could forgo watching one television program and instead stretch your mental muscles. The important thing is to make this time free from distractions. Reserve this time to do what you want to do to keep your mind in top form.

To help you guide your efforts, use the charts on the following pages. After two weeks, evaluate the results. Do the exercises seem to be doing what you want them to do? Do you find it easier to collect your attention? Can you concentrate more easily? Are your mental pictures becoming steadier? Do you notice things more? Are your perceptions more vivid?

If you feel more alert, then congratulations, the exercises are having an effect. If you don't feel a difference, or if you feel worse than before, then it's likely that you are not collecting your attention fully. Maybe these are not the best exercises for you, or there are other psychological factors at play.

Use the empty forms on pages 229 and 230 as guides to chart the progress of your formal exercise routine. Write down the exercises you did as well as how long you did them for. Add any thoughts, insights, or dilemmas that you may have experienced.

Remember to begin your exercise routine by loosening up. Relax, steady your breathing, and collect your attention. Work from the easy to the more difficult. If an exercise seems too hard, and you find yourself fighting your mind, then ease off, and do something else. Come back to it later. As you become more proficient, make the exercises more challenging. Devise your own mental exercises that test your mental fitness. Use the earlier exercises as a springboard to greater and more invigorating mental activity.

And remember, the whole idea of mental exercise is to explore your mind, so take the time to think. Don't skim. Get down to the root of your thinking. Penetrate the far corners of your mind which may be covered with the dust of everyday, routine thought.

*"Besides the noble art of getting things done,
master the noble art of leaving things undone.
The wisdom of life consists
in the elimination of nonessentials."*

LIN YUTANG, *Chinese mystic*

Week One

- [ ] **Day One:**   The Grand Tour
                            Two-Minute Mind

- [ ] **Day Two:**   The Grand Tour
                            Two-Minute Mind

- [ ] **Day Three:**   The Grand Tour
                            Two-Minute Mind

- [ ] **Day Four:**   Magic Massaging Finger
                            Variations on Two-Minute Mind

- [ ] **Day Five:**   Magic Massaging Finger
                            Number Exercises

- [ ] **Day Six:**   Magic Massaging Finger
                            Number Exercises

- [ ] **Day Seven:**   Attention Grabbers
                            Verse Exercises

Week Two

- ☐ **Day One:** Attention Grabbers
  Letter Exercises

- ☐ **Day Two:** Mental Manipulation
  Letter Exercises

- ☐ **Day Three:** Mental Manipulation
  Letter Exercises

- ☐ **Day Four:** Imagery Exercises
  Word Exercises

- ☐ **Day Five:** Imagery Exercises
  Exercises with a Verse

- ☐ **Day Six:** Imagery Exercises
  Number Exercises

- ☐ **Day Seven:** The Grand Tour
  Two-Minute Mind

Date: _____

Objective: _____

❑   **Day One:**   _____
                    _____
                    _____

❑   **Day Two:**   _____
                    _____
                    _____

❑   **Day Three:** _____
                    _____
                    _____

❑   **Day Four:**   _____
                    _____
                    _____

❑   **Day Five:**   _____
                    _____
                    _____

❑   **Day Six:**   _____
                    _____
                    _____

❑   **Day Seven:** _____
                    _____
                    _____

Date: _____

Objective: _____

❏ **Day One:** _____

_____

_____

❏ **Day Two:** _____

_____

_____

❏ **Day Three:** _____

_____

_____

❏ **Day Four:** _____

_____

_____

❏ **Day Five:** _____

_____

_____

❏ **Day Six:** _____

_____

_____

❏ **Day Seven:** _____

_____

_____

Where Do You Go from Here?

"If you take any activity,
any art, any discipline, any skill,
take it and push it as far as it will go,
push it beyond where it has ever been before,
push it to the wildest edge of edges,
then you force it into the realm of magic."

TOM ROBBINS, *author*

The mind is a remarkable thing. We can't touch it. We can't smell it. We can't hear it. We can't point to it. Yet it is always with us, making our world, and creating our lives.

Having a fit mind doesn't necessarily mean that you can learn differential calculus in an hour, or that you will never have bad moods again. It doesn't mean that you will always make the best decision, or that you will always find the right answer.

Having a fit mind means that you are able to think in the direction that you want to think. It's the ability to choose where you want your thoughts to go.

Where can you go with mental exercise? That's a question only you can answer. You may choose to explore the world of logical thought. You may choose to wander through the delicate images of creative imagination. You may wish to delve into words or numbers. You may want to explore the world of possibility or the world of practical application.

Eventually, you will discover little distinction between formal mental exercise and everyday thinking. Your mind will respond to the challenges that you give yourself. You will continue to learn by watching and by doing. Through the act of thinking actively, you will continue to keep your mind fit.

Keeping Fit Style Tips

Mental exercise stimulates your mind, massages your inner muscles, and refreshes your spirit. In the gymnasium of life, let your mental muscles do what they were meant to do: move.

TIP ONE Take responsibility for your mental condition. After all, if you're not going to take care of your brain, who will?

TIP TWO Encourage your mind to think actively rather than just to *have thoughts*. Decide what you want to think about and go ahead and think about it.

TIP THREE Practice mental exercise regularly. The more you exercise, the easier it becomes. The easier it becomes, the more you enjoy it. The more you enjoy, the better you'll feel. Make it a habit to challenge your mental muscles regularly. Don't just think about it. Do it.

"It is not enough to have a good mind.
The main thing is to use it well."
RENÉ DESCARTES, *17th-century mathematician and philosopher*

Return To
Mack's Mental Gymnasium

Tom's Return

Six weeks into his regular mental workout, Tom paid another visit to Mack. Mack looked Tom over and said, "You're looking much better. I see that in some places, you've taken off some extra mental weight, and in other places, you've filled out. How do you feel?"

"Very well, actually," Tom replied. "My attention is clearer. My thoughts are sharper. And my mind is more fit. I feel great."

"Terrific. Then what can I do for you?"

"Well, I don't want to lose the habit of exercising my mind. I know how easy it is to get into a rut and put off doing things that are good for me. Right now I'm doing fine, but I know that down the road my interest may lag, and that I'll probably fall back into my old mental habits. Is there anything that I can do to help keep my motivation up?"

Mack walked to the back of his office, reached into a locker, and pulled out a large black book. As Mack handed the book over, he said, "It's yours to keep. I've been keeping it until you were ready. To keep the habit of mental exercise alive, it's helpful to have some activity that you can repeat every day, which will encourage you to focus on your thinking. One of the best things to do is to keep a journal."

Tom asked, "What should I write in it?"

"You can use the journal in two ways. One way is to keep track of formal exercises you perform in the mental gymnasium. You can list which thinking routines, drills, visualizations, and meditations you did. Having a book gives you a solid reminder to exert your mental muscles. Over time, you build up an account of which exercises work best for you.

"The other way of using the journal is to use a technique that I call *freestyle*. Each evening, for a few minutes, you write down anything that you want to think about. You might write about how your day went. You might jot down notes about an insight that you had. You might write a treatise about your long-term plans. You might

use the time to sort out some decisions you need to make. You could use the time to ponder the deeper issues in life. Or you might play around and doodle.

"What *freestyle* does is give you an opportunity to let your mind decide what it needs to think about. After a few weeks of writing, the habit begins to take shape. You develop a momentum, and your natural intelligence has an opportunity every day to express itself. The journal is a symbol for the time you spend engaged in mental exercise.

"If at any time you're feeling mentally stiff, intellectually lethargic, or cerebrally weak, open up your exercise journal and start reading. As you find gems of thought buried in the pages, you'll begin to feel sharper.

"Mental exercise can be a means to help you sharpen the skills you need to deal effectively in the world. The exercises act like a springboard to launch you into better situations. But mental exercise can also become an end in itself. When properly done, the mental movement becomes a form of art, a marvel of beauty. When you're thinking in this way, thinking becomes an aesthetic pursuit."

Tom smiled and thanked Mack for the journal. As Tom went downstairs, he felt excited about his new prospect. He flipped through the empty pages of the journal and wondered what insights and thoughts he would eventually write down. Tom realized that he now had the means to help his thinking remain clear. It was simple, but then again the best things in life are.

Tom began to realize that he could choose to think in any direction he wanted. This gave him a great sense of freedom, as if someone had opened a window in his mind. He felt that he was becoming free from old habits that inhibited his mental agility.

Once outside, Tom had the strange feeling that, in a sense, the entire world was a mental gymnasium. As he looked at people's faces, the cracks in the concrete, as he listened to the sounds, as he thought about which direction he was going to walk, Tom realized that there were countless things to think about and to appreciate.

As he continued on his way, Tom looked towards the future, and thought to himself that he had begun the best workout of all.

END STUFF

Answers to Puzzles and Problems

Conditioning

PRESTIDIGITATION (Page 55) You can connect all the digits from one to nine in a meaningful equation with three signs in the following way: 123 - 45 - 67 + 89 = 100.

REVERSE PRESTIDIGITATION (Page 55) Make the digits nine to one equal one hundred in the following way: 98 - 76 + 54 + 3 + 21 = 100.

DOUBLETS (Page 60)
OAT, rat, rot, roe, RYE.
EYE, dye, die, did, LID.
HOT, sot, set, sea, TEA.
HALF, hale, tale, tile, TIME.
MOON, moan, mean, bean, BEAM.

PIG, wig, wag, way, say, STY.
REST, lest, lost, loft, soft, SOFA.
FISH, fist, gist, girt, gird, BIRD.
LIFE, lift, list, lest, best, BELT.
PORT, part, pant, pint, pine, WINE.

FAST, last, lost, loot, soot, slot, SLOW.
PAWN, paws, pans, pins, pink, kink, KING.
POOR, boor, book, rook, rock, rick, RICH.
FLOUR, floor, flood, blood, brood, broad, BREAD.
FOOT, fort, fore, fare, fate, pate, PATH.

NOSE, note, cote, core, corn, coin, CHIN.
PITCH, pinch, winch, wench, tench, tenth, TENTS.
PLAY, slay, slam, seam, ream, roam, ROOM.
COAL, coat, moat, most, mist, mint, MINE.
PEAR, peat, feat, feet, fret, free, TREE.

WHEAT, cheat, cheap, cheep, creep, creed, breed, BREAD.
FOUR, foul, fool, foot, fort, fore, fire, FIVE.
PITY, pits, pins, fins, find, fond, food, GOOD.

238

BLACK, blank, blink, clink, chink, chine, whine, WHITE.

STEAL, steel, steer, sheer, shier, shies, shins, chins, COINS.
WOOD, wold, weld, weed, feed, fled, flee, free, TREE.
ELM, ell, all, ail, air, fir, far, oar, OAK.
ARMY, arms, aims, dims, dams, dame, name, nave, NAVY.
BUY, bud, bid, aid, aim, arm, ark, ask, ASS.
ONE, owe, ewe, eye, dye, doe, toe, too, TWO.

ANAGRAMS (Page 61)

TREASON = SENATOR ; MEDUSA = AMUSED
DIRECTOR = CREDITOR ; CONSIDERATE = DESECRATION
WANDER = WARDEN ; MASCULINE = CALUMNIES
CATECHISM = SCHEMATIC ; EXCITATION = INTOXICATE
BESTIARY = SYBARITE ; DICTIONARY = INDICATORY
LEOPARD = PAROLED ; LEGISLATOR = ALLEGORIST

APPENDED ANAGRAMS (Page 61)

HAT, chat, catch.   LOUT, clout, occult.   HATE, cheat, cachet.
SEAR, scare, scarce.   TAPE, epact, accept.   ARK, rack, crack.
SPITE, septic, sceptic. HEAD, ached, cached.   HERE, cheer, creche.
RILE, relic, circle.   IRK, rick, crick.   NOSE, scone, sconce.
LEAN, lance, cancel.   NEAR, crane, cancer.   NEAT, enact, accent.

ALPHABET ADVANCE (Page 61)

TIGER = PECAN (four letters back).
JOLLY = CHEER (seven letters back).
ADDER = BEEFS (one letter forward).
FILLS = LORRY (six letters forward).
SORRY = MILLS (six letters back).
FREUD = COBRA (three letters back).
CHAIN = INGOT (six letters forward).
FERNS = BANJO (four letters back).

Words

CIRCLE OF THOUGHT (Page 96) There are at least twenty words in the circle.
UMBRA, ANGER, GERANIUM, RAN, RAG, BRAG, AGE, AGED, ERA, DRIES, DRIEST, RANGE, RAGE, RAGED, UMBRAGE, RANGER, STRANGER, ESTRANGE, AN, RANG

Problem Solving

TALLAHASEE ZOO (Page 109) If (Beasts + Birds = 30) then (Beasts = 30 - Birds). Into the equation (4 x Beasts + 2 x Birds = 100) substitute (Beasts = 30 - Birds) giving (4 x (30 - Birds)) + (2 x Birds) = 100. Therefore 120 - (4 x Birds) + (2 x Birds) = 100. 120 - 2 x Birds = 100. 20 = 2 x Birds. There are ten birds and twenty beasts in the zoo.

THREE MISSIONARIES AND THREE CANNIBALS (Page 109)
First a missionary and a cannibal cross. The missionary returns.
Two cannibals cross. One cannibal returns.
Two missionaries cross. A missionary and a cannibal return.
Two missionaries cross. One cannibal returns.
Two cannibals cross. One cannibal returns.
The remaining two cannibals cross.

WORD ADDITIONS (Page 110)
SEVEN - NINE = EIGHT. There are two solutions.

```
  2 1 5 1 4            5 4 1 4 6
- _ 4 6 4 1_         - _6 7 6 4_
  1 6 8 7 3            4 7 3 8 2
```

SEND MORE MONEY   ABCDE X 4 = EDCBA

```
    9 5 6 7              2 1 9 7 8
+ _ 1 0 8 5_          x _      4_
  1 0 6 5 2              8 7 9 1 2
```

WIRE MORE MONEY

```
   9 7 6 2
+  1 0 6 2
 1 0 8 2 4
```

WRONG + WRONG = RIGHT

```
   2 5 9 3 8
+  2 5 9 3 8
   5 1 8 7 6
```

LETTERS + ALPHABET = SCRABBLE

```
    7 0 8 8 0 6 2
+ 1 7 5 3 1 9 0 8
  2 4 6 1 9 9 7 0
```

FLIP IT OR DIP IT (Page 110) Work backwards to solve this problem. Begin with the amount of money the players have after the third game, then halve two players, money and double the third player's money. This gives you the money after two games. Do this two more times.

	Player One	Player Two	Player Three
3rd game	$36.00	$36.00	$36.00
2nd game	$18.00	$18.00	$72.00
1st game	$9.00	$63.00	$36.00
Start	$58.50	$31.50	$18.00

EARRINGS (Page 110) There are 500 earrings in the village.

STAN AND OLLIE (Page 110) The proportions are the same. Say that a teaspoon contains ten percent of the cup's total volume. A teaspoon of Stan's coffee – milk mixture would then contain ninety percent coffee and ten percent milk. That means 9 / 10 of a teaspoon of coffee is introduced to the milk. But since the spoon contains 1 / 10 of a teaspoon of milk, the remaining 9 / 10 teaspoon of milk remains in the coffee. Each cup contains 9 / 10 of a teaspoon of the other liquid.

GRAPHIC ARTIST (Page 112) Arrange the statues in this way.

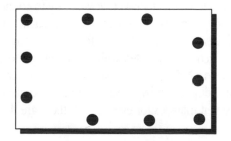

ONE LINE PUZZLE (Page 112) The solution to the two one-liners is as follows:

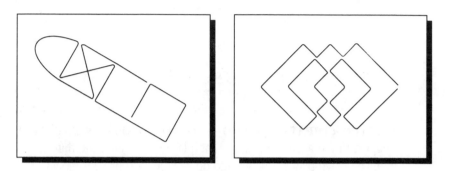

MAGIC HEXAGON (Page 112) First fold the paper in half, horizontally and vertically to form the lines AB and CD. Then fold A to O to get GE and B to O to get FH. Fold over AJ so that J lies on the line GE — at the point G. Do the same for the other three corners to get the points EFH. Then it is straightforward to find the hexagon.

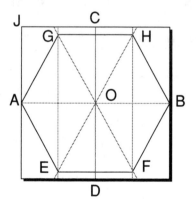

PENTAGRAM (Page 113) There are thirty-five triangles in the pentagram. To find the correct answer, you can try to count the number of triangles, or you can take a systematic approach. Since the figure is symmetrical, it's possible to select a reference point and count the number of triangles which are associated with that point — those which either join or are opposed to it. For the top point, there is a pattern of seven triangles which are not duplicated by selecting another point. Since there are five points and seven triangles for every point, there are thirty five-triangles.

SEVEN MEN AND TWO BOYS (Page 113) This puzzle can be put into the form of a pattern. First, two boys cross. One boy returns. A boy and a man cross. Two boys return. So it takes four trips to carry one man across the river. Since there are seven men, twenty-eight crossings are needed.

MOUSE MAZE (Page 113) The total number of routes is thirty-five. The problem can be solved with the realization that the number of routes to a particular square equals the sum of the number of routes of the two squares that lead into the first square. Use the diagram to add up the possible routes.

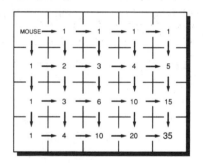

CIRCLE ADDITION (Page 114) Here is one of many arrangements of numbers that lead to a solution.

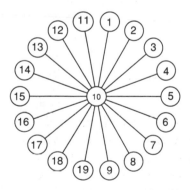

RADAR (Page 114) Every word must begin with the letter R. There are twenty possible ways of reading RADAR for each R. Since there are four R's in the figure, there is a total of eighty ways to read the word.

COOKIE JARS (Page 116) Take a cookie from peanut butter - oatmeal jar. Since the jars are labeled incorrectly, you will get either an oatmeal or a peanut butter cookie. If you get an oatmeal cookie, you label the peanut butter - oatmeal jar as oatmeal. What was labeled the peanut butter jar must contain peanut butter - oatmeal cookies because we are told that the jars are all incorrectly labeled. What was originally labeled as oatmeal jar must contain peanut butter cookies.

CHAIN LINK (Page 116) If you cut five links from one chain and use these to connect the others you only need to spend $(4 \times 5) + (4 \times 7) = 48$ cents.

SOCK IT TO ME (Page 116) You only need to remove three socks to ensure that you have a matching pair.

SILVER (Page 116) The house owner can cut the silver bar with three slices, to make four pieces measuring one inch, two inches, four inches, and eight inches. On day one she gives the worker the smallest piece. One day two, she takes back the one-inch piece and gives him two inches of silver. On day three, she gives him one inch more. On day four, she takes back the one-inch and two-inch pieces and gives him the four-inch piece, and so on.

UNCLE JAKE'S COINS (Page 116) First, weigh sixteen coins with eight on each side of the balance. If one side is heavier, that side contains the heavier coin. If the sides are equal in weight, then the heavier coin is in the group that you did not weigh. Second, from the group that contains the gold coin, take six coins and compare their weight. If one side is heavier, you know that group of three coins contains the gold coin. If the sides are equal, the remaining group of two contains the gold coin. Third, weigh the final group. If the gold coin is in a group of three, weight two coins. If the balance is tipped, you have found the heavier coin. If the balance is level, the odd coin is made of gold.

FORK IN THE ROAD (Page 116) You can ask either guard the following question: "If I were to ask you if this was the correct route to Tipperary, would you say yes?" At the same time, you should point along the road.

SIX MATCHES (Page 119) The solution is to construct a tetrahedron.

FOUR MATCHES (Page 119) The solution is to move the bottom match leaving a tiny square in the center of the figure.

SILK ROPES (Page 119) The thief found a solution by tying the end of the ropes together. He scaled one of the ropes to the top. He cut the other rope about a foot from the top and let the rest fall. From this one - foot piece, he tied a loop. Then he climbed to the loop, cut the remaining rope and threaded it through the loop. Then by holding both ends of the rope he lowered himself down.

JANE'S CAB (Page 120) If the cabbie was deaf, how did he know where to take her? As well, if the cabbie could not hear, why did he say that he couldn't hear a word of what she was saying?

EVIL MONEYLENDER (Page 120) The woman quickly reached into the bag, pulled out the pebble and nervously dropped it before anyone had a chance to see its color. The pebble was hopelessly lost among the other pebbles. After a moment, she reached into the bag again and removed the remaining pebble. Since it was black, the first must have been white. By turning the trick against the moneylender, the young woman was free.

PORTHOLE RISING (Page 120) The water never reaches the port- hole because the ship rises with the water.

PASSING TRAINS (Page 120) The trains passed through the tunnel at different times of the afternoon.

INFLATION (Page 120) The answer is that 1988 dollar bills is one more than 1987 dollar bills.

BACTERIA (Page 121) The container is a quarter full four hours before midnight, at eight o'clock.

SUDDEN DEPARTURE (Page 121) The wife was on a life support system. When the man pressed the elevator button, nothing happened, in- dicating that the power had failed, turning off her life support system.

ALBRATROSS (Page 121) The sailor is blind. He was once ship-wrecked and, in order to live, he and his co-survivors, ate what he was told was albatross. Since the albatross in the restaurant did not taste like the albatross he ate while marooned, the man concluded that the first albatross was the flesh of his dead crewmates.

MURDER SUPECT (121) The woman is a Siamese twin.

STRANGE BEHAVIOUR (121) The woman had hiccups.

PENNY PUZZLE (Page 121) Fold the paper across the hole. Drop in a quarter, and bend the paper so that opening becomes wider.

SIMPLY AMAZING (Page 133) Here is the solution to the maze.

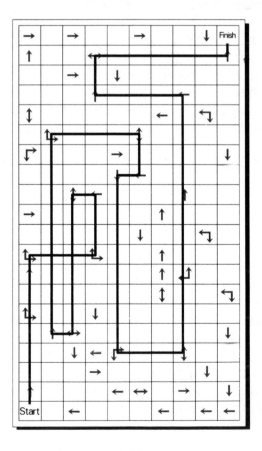

NUMBER PATTERNS (Page 166)   The next letter in the sequence A E F H I K L M ... is N. Each letter is composed only of straight lines. D31. The letters stand for the months of the year, and the numbers tell the number of days in that month.

The number is 10. Successive numbers are generated by adding the digits of the previous number and multiplying by two.

The next number is 5. This sequence is the digits in the number pi.

The next letter in the sequence Q, W, E, R, T, ... is Y. These letters form the sequence on a typewriter keyboard.

HUEY, DEWEY, AND LEWEY (Page 167)   Harry, Huey and Dewey, Lewey, Dick, Tom

TOM, DICK, AND HARRY (Page 167)
Tom is the oldest, Dick is the next oldest, and Harry is the youngest. A good way to solve this problem is to put it into mathematical notation. Tom is twice as old as Dick will be implies that $T > D$. The statement when Harry is as old as Tom is now also implies that $T > H$. Therefore Tom is the oldest. When Harry equals Tom's current age, Tom will be twice as old as Dick can be restated when $T = H$, $T = 2D$.

MENTAL OLYMPICS (Page 167)   Since everyone put up his hand, Harry figured there must be two possibilities: either two or three people are wearing a red hat. Harry also figured that if anyone could *see* a white hat, they could deduce the color of their own hat as being red. Since no one did this, Harry realized that nobody saw a white hat. Therefore everyone was wearing a red hat. Therefore Harry had a red hat.

HEN (Page 167)   The answer is one hen. If a hen and a half can lay an egg and a half in one and half days, then one hen can lay one egg in one and a half days. A hen that lays better by a half can therefore lay one and a half eggs in a day and a half, or one egg in a day. In ten and a half days (a week and a half) that hen can lay ten and a half eggs (half a score and a half).

About The Author

Tom Wujec is currently a writer, producer, and lecturer at the McLaughlin Planetarium in Toronto, Canada. He has a degree in Astronomy and Psychology from the University of Toronto.

Tom has traveled around the world twice and has explored India, Nepal, and Thailand. He has worked as a photographer, a drywaller, a vacuum cleaner salesman, and a Bunraku puppeteer. Tom also has an extensive background in computer graphics, audio visual communications, and interactive videodisk technology. He also runs a computer consulting business which specializes in computer aided design, desktop publishing and Postscript programing.

Besides traveling and writing, his passions include Go, Tai Chi, cycling, and, if he can just get the knack, juggling.

I hope that you have enjoyed reading this book and performing the mental exercises. If you have any thoughts or experiences you'd like to pass along, please feel free to drop a line. I'd be delighted to hear from you. Address all correspondence to

Tom Wujec
Doubleday Canada
105 Bond Street
Toronto, Ontario
Canada
M5B 1Y3